CliffsNotes®

Basic Math and Pre-Algebra

D0730988

CD includes 600 practice problems

Practice Pack

Basic Math and Pre-Algebra

CD includes 600 practice problems

Practice Pack

By Jonathan J. White

Contributing Authors
Teri Stimmel
Scott Searcy
Danielle Lutz

Wiley Publishing, Inc.

Published by:
Wiley Publishing, Inc.
111 River Street
Hoboken, NJ 07030-5774
www.wiley.com

Copyright © 2010 Wiley Publishing, Inc., Hoboken, New Jersey

Published by Wiley Publishing, Inc., Hoboken, New Jersey
Published simultaneously in Canada

Library of Congress Control Number: 2009933751

ISBN: 978-0-470-53349-9

Printed in the United States of America

10 9 8 7 6 5 4 3 2 1

No part of this publication may be reproduced, stored in a retrieval system, or transmitted in any form or by any means, electronic, mechanical, photocopying, recording, scanning, or otherwise, except as permitted under Sections 107 or 108 of the 1976 United States Copyright Act, without either the prior written permission of the Publisher, or authorization through payment of the appropriate per-copy fee to the Copyright Clearance Center, 222 Rosewood Drive, Danvers, MA 01923, 978-750-8400, fax 978-646-8600, or on the web at www.copyright.com. Requests to the Publisher for permission should be addressed to the Permissions Department, John Wiley & Sons, Inc., 111 River Street, Hoboken, NJ 07030, (201) 748-6011, fax (201) 748-6008, or online at http://www.wiley.com/go/permissions.

The publisher and the author make no representations or warranties with respect to the accuracy or completeness of the contents of this work and specifically disclaim all warranties, including without limitation warranties of fitness for a particular purpose. No warranty may be created or extended by sales or promotional materials. The advice and strategies contained herein may not be suitable for every situation. This work is sold with the understanding that the publisher is not engaged in rendering legal, accounting, or other professional services. If professional assistance is required, the services of a competent professional person should be sought. Neither the publisher nor the author shall be liable for damages arising herefrom. The fact that an organization or Website is referred to in this work as a citation and/or a potential source of further information does not mean that the author or the publisher endorses the information the organization or Website may provide or recommendations it may make. Further, readers should be aware that Internet Websites listed in this work may have changed or disappeared between when this work was written and when it is read.

Trademarks: Wiley, the Wiley Publishing logo, CliffsNotes, the CliffsNotes logo, Cliffs, CliffsAP, CliffsComplete, CliffsQuickReview, CliffsStudySolver, CliffsTestPrep, CliffsNote-a-Day, cliffsnotes.com, and all related trademarks, logos, and trade dress are trademarks or registered trademarks of John Wiley & Sons, Inc. and/or its affiliates. All other trademarks are the property of their respective owners. Wiley Publishing, Inc. is not associated with any product or vendor mentioned in this book.

For general information on our other products and services or to obtain technical support, please contact our Customer Care Department within the U.S. at (877) 762-2974, outside the U.S. at (317) 572-3993, or fax (317) 572-4002.

Wiley also publishes its books in a variety of electronic formats. Some content that appears in print may not be available in electronic books. For more information about Wiley products, please visit our web site at www.wiley.com.

Note: If you purchased this book without a cover, you should be aware that this book is stolen property. It was reported as "unsold and destroyed" to the publisher, and neither the author nor the publisher has received any payment for this "stripped book."

WILEY is a trademark of Wiley Publishing, Inc.

About the Authors

Jonathan J. White is Assistant Professor of Mathematics at Coe College in Cedar Rapids, Iowa. **Scott Searcy,** a high school teacher in Iowa, holds a Bachelor of Arts degree in Math and General Science. **Teri Stimmel** graduated from Coe College with a Bachelor of Arts degree in Mathematical Science and Secondary Education. **Danielle Lutz** is a graduate student at Coe College.

Publisher's Acknowledgments

Editorial

Project Editor: Donna Wright

Acquisitions Editor: Greg Tubach

Technical Editor: Tom Page

Composition

Project Coordinator: Kristie Rees

Indexer: Broccoli Information Mgt.

Proofreader: Susan Hobbs

Wiley Publishing, Inc., Composition Services

Table of Contents

Pretest

Problems

1. Which of the following are natural numbers?

   A. −3

   B. $\frac{1}{4}$

   C. $\sqrt{2}$

   D. 5

   E. Exactly two of the above are natural numbers.

   F. None of the above are natural numbers.

2. Which of the following are integers?

   A. $\frac{-7}{3}$

   B. −6

   C. π

   D. 10

   E. $\sqrt[3]{7}$

   F. Exactly two of the above are integers.

   G. None of the above are integers.

3. Which of the following are rational numbers?

   A. $\sqrt[3]{3}$

   B. $\frac{\pi}{2}$

   C. $\frac{-3}{5}$

   D. $\sqrt{5}$

   E. 5

   F. Exactly two of the above are rational numbers.

   G. None of the above are rational numbers.

4. Simplify $4[30 - 5(3 - 2)]$.

   A. 20

   B. 100

   C. 220

   D. 95

   E. 103

   F. None of the above

5. Simplify $4^2 \times 2 + 12 \div 2$.
   A. 38
   B. 44
   C. 22
   D. 14
   E. 3
   F. None of the above

6. Simplify $20 \div 2^2 + 3 \times 2 - (8 - 3)$.
   A. 101
   B. 11
   C. 5
   D. 0
   E. 6
   F. None of the above

7. Add $3 + 17 + 22$.
   A. 42
   B. 32
   C. 69
   D. 15
   E. 17
   F. None of the above

8. Add $32 + 14 + 19$.
   A. 55
   B. 14
   C. 28
   D. 65
   E. 42
   F. None of the above

9. Add $834 + 128$.
   A. 1052
   B. 952
   C. 962
   D. 902
   E. 862
   F. None of the above

10. Subtract 35 − 14.

    A. 11

    B. 21

    C. 26

    D. 31

    E. 14

    F. None of the above

11. Subtract 134 − 28.

    A. 94

    B. 124

    C. 116

    D. 16

    E. 106

    F. None of the above

12. Subtract 450 − 193.

    A. 257

    B. 167

    C. 267

    D. 357

    E. 67

    F. None of the above

13. Multiply 27×9.

    A. 153

    B. 36

    C. 136

    D. 243

    E. 216

    F. None of the above

14. Multiply 73×14.

    A. 365

    B. 1022

    C. 2993

    D. 2011

    E. 1752

    F. None of the above

15. Multiply 533 × 22.
    A. 11,726
    B. 1166
    C. 1826
    D. 242
    E. 1066
    F. None of the above

16. Divide 114 ÷ 3.
    A. 28
    B. 38
    C. $4\frac{2}{3}$
    D. 14
    E. None of the above

17. Divide 350 ÷ 25.
    A. 8
    B. 10
    C. 12
    D. 14
    E. None of the above

18. Divide 728 ÷ 13.
    A. 66
    B. 46
    C. 52
    D. 62
    E. None of the above

19. Round 5635 to the nearest ten.
    A. 5630
    B. 5635
    C. 5600
    D. 5640
    E. None of the above

20. Round 545 to the nearest hundred.

    A. 500
    B. 600
    C. 550
    D. 1000
    E. None of the above

21. Round 67,891 to the nearest thousand.

    A. 67,000
    B. 67,890
    C. 68,000
    D. 67,900
    E. None of the above

22. What are the factors of 20?

    A. 4 and 5
    B. 2 and 10
    C. 1, 2, 5, and 10
    D. 1, 2, 4, 5, 10, and 20
    E. None of the above

23. Which of the following numbers is prime?

    A. 6
    B. 10
    C. 7
    D. 2
    E. 25
    F. Exactly two of the above are prime.
    G. None of the above are prime.

24. Which of the following numbers are composite?

    A. 45
    B. 11
    C. 12
    D. 19
    E. 3
    F. Exactly two of the above are composite.
    G. None of the above are composite.

25. Add $\frac{1}{3} + \frac{3}{4}$.

    A. $\frac{4}{7}$

    B. $\frac{1}{4}$

    C. $\frac{13}{12}$

    D. $\frac{1}{3}$

    E. None of the above

26. Add $\frac{5}{8} + \frac{2}{3}$.

    A. $\frac{7}{11}$

    B. $\frac{31}{24}$

    C. $\frac{7}{24}$

    D. $\frac{23}{12}$

    E. None of the above

27. Add $\frac{1}{8} + \frac{3}{5}$.

    A. $\frac{29}{40}$

    B. $\frac{4}{13}$

    C. $\frac{1}{10}$

    D. $\frac{1}{5}$

    E. None of the above

28. Subtract $\frac{7}{9} - \frac{2}{3}$.

    A. $\frac{5}{6}$

    B. $\frac{14}{9}$

    C. $\frac{14}{27}$

    D. $\frac{1}{9}$

    E. None of the above

29. Subtract $\frac{5}{6} - \frac{2}{7}$.

    A. $\frac{5}{21}$

    B. 3

    C. $\frac{1}{14}$

    D. $\frac{1}{2}$

    E. None of the above

30. Subtract $\frac{1}{3} - \frac{1}{5}$.

    A. 0

    B. $\frac{2}{15}$

    C. $\frac{1}{35}$

    D. $\frac{1}{15}$

    E. None of the above

31. Add $2\frac{1}{3} + 3\frac{1}{5}$.

    A. $5\frac{1}{35}$

    B. $5\frac{1}{15}$

    C. $5\frac{1}{4}$

    D. $5\frac{8}{15}$

    E. None of the above

32. Subtract $3\frac{1}{3} - 2\frac{1}{6}$.

    A. $1\frac{1}{6}$

    B. $\frac{1}{36}$

    C. $2\frac{1}{6}$

    D. $5\frac{1}{2}$

    E. None of the above

33. Subtract $4\frac{1}{2} - 2\frac{2}{3}$.

    A. $\frac{6}{11}$

    B. $\frac{5}{6}$

    C. $2\frac{5}{6}$

    D. 8

    E. None of the above

34. Multiply $\frac{4}{3} \times \frac{2}{13}$.

    A. 8

    B. $\frac{8}{39}$

    C. $\frac{26}{3}$

    D. $\frac{3}{8}$

    E. None of the above

35. Multiply $\frac{5}{6} \times \frac{3}{10}$.
    A. $\frac{1}{2}$
    B. $\frac{3}{2}$
    C. $\frac{25}{9}$
    D. $\frac{2}{15}$
    E. None of the above

36. Multiply $3\frac{1}{2} \times 2\frac{2}{3}$.
    A. $\frac{28}{3}$
    B. $6\frac{1}{6}$
    C. 2
    D. $\frac{5}{6}$
    E. None of the above

37. Divide $\frac{3}{5} \div \frac{1}{10}$.
    A. $\frac{3}{15}$
    B. $\frac{1}{2}$
    C. 6
    D. $\frac{3}{50}$
    E. None of the above

38. Divide $1\frac{3}{5} \div \frac{2}{3}$.
    A. $\frac{5}{12}$
    B. $\frac{14}{15}$
    C. $\frac{16}{15}$
    D. $\frac{1}{2}$
    E. None of the above

39. Divide $1\frac{1}{2} \div 3\frac{1}{3}$.
    A. 3
    B. $\frac{9}{20}$
    C. 5
    D. $\frac{29}{6}$
    E. None of the above

40. Add 3.2 + 7.8.

    A. 10
    B. 3
    C. 12
    D. 4.6
    E. None of the above

41. Add 2.25 + 1.5.

    A. 3.3
    B. 2.4
    C. 3.75
    D. 0.75
    E. None of the above

42. Add 48.3 + 3.62.

    A. 84.5
    B. 8.45
    C. 0.845
    D. 51.92
    E. None of the above

43. Subtract 7.6 − 3.3.

    A. 0.43
    B. 4.3
    C. 3.3
    D. 56.3
    E. None of the above

44. Subtract 90.3 − 8.4.

    A. 9.9
    B. 81.9
    C. 82.9
    D. 1.9
    E. None of the above

45. Subtract 462.3 − 100.5.

    A. 362.8

    B. 562.8

    C. 451.8

    D. 388.0

    E. None of the above

46. Multiply 2.3 × 4.5.

    A. 6.8

    B. 103.5

    C. 10.35

    D. 2.25

    E. None of the above

47. Multiply 15.3 × 0.003.

    A. 0.459

    B. 4.59

    C. 459.0000

    D. 15.303

    E. None of the above

48. Multiply 60.1 × 6.1.

    A. 37.21

    B. 372.1

    C. 66.2

    D. 366.61

    E. None of the above

49. Divide 16.5 ÷ 0.5.

    A. 33

    B. 3300

    C. 3.2

    D. 11.2

    E. None of the above

50. Divide 14.7 ÷ 0.25.

    A. 5.88

    B. 3.675

    C. 588

    D. 1.7

    E. None of the above

51. Divide $2.3 \div 0.6$.

   A. 1.7

   B. 3

   C. $38.\overline{3}$

   D. $3.8\overline{3}$

   E. None of the above

52. Change 0.45 to a fraction.

   A. $4\frac{1}{2}$

   B. 45%

   C. $\frac{5}{4}$

   D. $\frac{9}{20}$

   E. None of the above

53. Change $\frac{2}{9}$ to a decimal.

   A. 0.29

   B. 9.2

   C. 45%

   D. $0.\overline{2}$

   E. None of the above

54. Change $2.\overline{3}$ to a fraction.

   A. $\frac{2}{3}$

   B. $\frac{7}{3}$

   C. $\frac{1}{3}$

   D. $2.\overline{3}$ is irrational and cannot be written as a fraction.

   E. None of the above

55. Change 0.20 to a percent.

   A. 2%

   B. 20%

   C. 5%

   D. $\frac{1}{2}$%

   E. None of the above

56. Change 75% to a fraction.

    A. $\frac{1}{4}$

    B. $\frac{1}{75}$

    C. $\frac{4}{3}$

    D. $\frac{75}{1}$

    E. None of the above

57. Change $\frac{3}{5}$ to a percent.

    A. 35%

    B. 53%

    C. 3.5%

    D. 60%

    E. None of the above

58. What is 25% of 90?

    A. 25

    B. 9.5

    C. 52

    D. 18

    E. None of the above

59. 24 is what percent of 30?

    A. 50%

    B. 24%

    C. 30%

    D. 80%

    E. None of the above

60. 45 is 25% of what number?

    A. 180

    B. 80

    C. 11.25

    D. 7

    E. None of the above

61. Add −5 + −3.

    A. 8

    B. −8

    C. 2

    D. −2

    E. None of the above

62. Subtract −3 − +5.

    A. 8

    B. −8

    C. 2

    D. −2

    E. None of the above

63. Add −11 + 5.

    A. 16

    B. −16

    C. −6

    D. 6

    E. None of the above

64. Multiply (−3) × (7).

    A. −10

    B. 10

    C. −21

    D. 21

    E. None of the above

65. Divide (−4) ÷ (−2).

    A. 8

    B. −8

    C. 2

    D. −2

    E. None of the above

66. Multiply $(-4) \times (2) \times (-3) \times (-1)$.

   A. -10

   B. -24

   C. 10

   D. 24

   E. None of the above

67. $|-5|$ is equal to

   A. 25

   B. $\frac{1}{5}$

   C. 5

   D. -5

   E. None of the above

68. $5 - |-6|$ is equal to

   A. 11

   B. -1

   C. 1

   D. $\frac{5}{6}$

   E. None of the above

69. $|15 - 18|$ is equal to

   A. 33

   B. -3

   C. -33

   D. 3

   E. None of the above

70. Add $\frac{1}{6} + \frac{-1}{2}$.

   A. $\frac{2}{3}$

   B. $\frac{-2}{3}$

   C. 0

   D. $\frac{-1}{3}$

   E. None of the above

71. Subtract $\frac{-2}{3} - \frac{-1}{4}$.

    A. $\frac{-5}{12}$

    B. $\frac{-11}{12}$

    C. $\frac{5}{12}$

    D. $\frac{11}{12}$

    E. None of the above

72. Add $\frac{2}{3} + \frac{-1}{4}$.

    A. $\frac{-5}{12}$

    B. $\frac{-11}{12}$

    C. $\frac{5}{12}$

    D. $\frac{11}{12}$

    E. None of the above

73. Multiply $\frac{2}{5} \times \frac{-1}{6}$.

    A. $\frac{1}{15}$

    B. $\frac{-1}{15}$

    C. $\frac{2}{11}$

    D. $\frac{-2}{11}$

    E. None of the above

74. Divide $\frac{-2}{5} \div \frac{-5}{3}$.

    A. $\frac{2}{3}$

    B. $\frac{-2}{3}$

    C. $\frac{6}{25}$

    D. $\frac{-6}{25}$

    E. None of the above

75. Divide $\frac{3}{5} \div \frac{-4}{15}$.

    A. $\frac{4}{9}$

    B. $\frac{-4}{9}$

    C. $\frac{4}{25}$

    D. $\frac{-4}{25}$

    E. None of the above

76. Expand 2^4.

   A. 8

   B. 6

   C. 16

   D. 64

   E. None of the above

77. Expand 3^{-2}.

   A. $\frac{1}{9}$

   B. -6

   C. -9

   D. 9

   E. None of the above

78. Expand 4^0.

   A. 0

   B. 4

   C. $\frac{1}{4}$

   D. 1

   E. None of the above

79. Multiply $5^{-1} \times 5^3$.

   A. 5^{-3}

   B. 5^2

   C. 5^{-4}

   D. 25^{-3}

   E. None of the above

80. Divide $8^3 \div 8^{-2}$.

   A. 8^5

   B. 8^1

   C. 1^1

   D. 0

   E. None of the above

81. Simplify $(10^3)^4$.

   A. 10^4

   B. 10^3

   C. 10^7

   D. 10^{12}

   E. None of the above

82. Simplify $\sqrt{9}$.

    A. 81

    B. 0

    C. 18

    D. This cannot be done with real numbers.

    E. None of the above

83. Simplify $\sqrt[3]{27}$.

    A. 3

    B. 8

    C. 9

    D. 1

    E. None of the above

84. Simplify $\sqrt{98}$.

    A. 49

    B. 7

    C. 10

    D. $7\sqrt{2}$

    E. None of the above

85. Express 4,500,000 in scientific notation.

    A. 4.5×10^6

    B. 4.5×10^{-6}

    C. 0.45×10^5

    D. 45×10^7

    E. None of the above

86. Express 0.00034 in scientific notation.

    A. 34×10^3

    B. 3.4×10^{-4}

    C. 43×10^{-1}

    D. 0.00034×10^{14}

    E. None of the above

87. Multiply 500×0.0012 and express the answer in scientific notation.

    A. 6×10^6

    B. 4.5×10^3

    C. 6×10^{-1}

    D. 1.2×10^{-3}

    E. None of the above

88. If a rectangle has sides of length 5 inches and 6 inches, then its area is

A. 9 square inches

B. 15 square inches

C. 22 square inches

D. 30 square inches

E. None of the above

89. If a triangle has a base of 10 centimeters, a height of 12 centimeters, and another side with a length of 6 centimeters, then its area is

A. 60 square centimeters

B. 120 square centimeters

C. 28 square centimeters

D. 22 square centimeters

E. None of the above

90. If a circle has a radius of 6 feet, then its circumference is

A. 36π feet

B. 144π feet

C. 6π feet

D. 12π feet

E. None of the above

91. If a blue marble, a red marble, and a white marble are placed in a jar and one is drawn at random, what is the probability that the red marble is the one drawn?

A. $\frac{1}{3}$

B. 2

C. 100

D. $\frac{1}{2}$

E. None of the above

92. If a standard die is tossed twice, what is the probability that the sum of the two numbers rolled will be an 11?

A. 1

B. $\frac{1}{2}$

C. $\frac{1}{18}$

D. 12

E. None of the above

93. If a card is drawn at random from a standard deck (excluding jokers), what is the probability that the card drawn is a king or queen?

    A. $\frac{1}{26}$

    B. 10

    C. $\frac{1}{2}$

    D. More than 10

    E. None of the above

94. Give an algebraic expression for five less than a number (use x to represent the number).

    A. $5 + x$

    B. $x + 5$

    C. $5 - x$

    D. $x - 5$

    E. None of the above

95. Give an algebraic expression for twice a number (use y to represent the number).

    A. $2y$

    B. $y + 2$

    C. $y - 2$

    D. $\frac{y}{2}$

    E. None of the above

96. Give an algebraic expression for seven more than a number (use a to represent the number).

    A. $a - 7$

    B. $a + 7$

    C. $a \times 7$

    D. $a \div 7$

    E. None of the above

97. Solve $z - 4 = 10$ for z.

    A. $z = 6$

    B. $z = 2.5$

    C. $z = 14$

    D. $z = 40$

    E. None of the above

98. Solve $4x + 3 = 15$ for x.

   A. $x = 3$

   B. $x = 4$

   C. $x = 4.5$

   D. $x = 15$

   E. None of the above

99. Solve $\frac{y}{3} - 5 = 1$ for y.

   A. $y = 8$

   B. $y = 18$

   C. $y = 6$

   D. $y = 2$

   E. None of the above

Answers

1. D

2. F

3. F

If you missed two out of the previous three questions, begin your studying by turning to page 27, Collections of Numbers.

4. B

5. A

6. E

If you missed two out of the previous three questions, begin your studying by turning to page 31, Grouping Symbols and Order of Operations.

7. A

8. D

9. C

If you missed two out of the previous three questions, begin your studying by turning to page 37, Adding and Subtracting.

10. B

11. E

12. A

If you missed two out of the previous three questions, begin your studying by turning to page 37, Adding and Subtracting.

13. D

14. B

15. A

If you missed two out of the previous three questions, begin your studying by turning to page 39, Multiplying.

16. B

17. D

18. E

If you missed two out of the previous three questions, begin your studying by turning to page 39, Multiplying.

19. D

20. A

21. C

If you missed two out of the previous three questions, begin your studying by turning to page 45, Rounding Off.

22. D

23. F

24. F

If you missed two out of the previous three questions, begin your studying by turning to page 50, Factoring.

25. C

26. B

27. A

If you missed two out of the previous three questions, begin your studying by turning to page 62, Adding and Subtracting Fractions.

28. D

29. E

30. B

If you missed two out of the previous three questions, begin your studying by turning to page 62, Adding and Subtracting Fractions.

31. D

32. A

33. E

If you missed two out of the previous three questions, begin your studying by turning to page 65, Adding and Subtracting Mixed Numbers.

34. B

35. E

36. A

If you missed two out of the previous three questions, begin your studying by turning to page 68, Multiplying Fractions and Mixed Numbers.

37. C

38. E

39. B

If you missed two out of the previous three questions, begin your studying by turning to page 71, Dividing Fractions and Mixed Numbers.

40. E

41. C

42. D

If you missed two out of the previous three questions, begin your studying by turning to page 78, Adding Decimals.

43. B

44. B

45. E

If you missed two out of the previous three questions, begin your studying by turning to page 79, Subtracting Decimals.

46. C

47. E

48. D

If you missed two out of the previous three questions, begin your studying by turning to page 81, Multiplying Decimals.

49.   A

50.   E

51.   D

If you missed two out of the previous three questions, begin your studying by turning to page 84, Dividing Decimals.

52.   D

53.   D

54.   B

If you missed two out of the previous three questions, begin your studying by turning to page 90, Changing between Fractions and Decimals.

55.   B

56.   E

57.   D

If you missed two out of the previous three questions, begin your studying by turning to page 97, Changing Percents, Decimals, and Fractions.

58.   E

59.   D

60.   A

If you missed two out of the previous three questions, begin your studying by turning to page 101, Applications of Percents.

61.   B

62.   B

63.   C

If you missed two out of the previous three questions, begin your studying by turning to page 113, Adding Integers or page 115, Subtracting Integers.

64.   C

65.   C

66.   B

If you missed two out of the previous three questions, begin your studying by turning to page 118, Multiplying and Dividing Integers.

67. C

68. B

69. D

If you missed two out of the previous three questions, begin your studying by turning to page 120, Absolute Value.

70. D

71. A

72. C

If you missed two out of the previous three questions, begin your studying by turning to page 122, Adding Positive and Negative Fractions or page 126, Subtracting Positive and Negative Fractions.

73. B

74. C

75. E

If you missed two out of the previous three questions, begin your studying by turning to page 128, Multiplying Positive and Negative Fractions or page 131, Dividing Positive and Negative Fractions.

76. C

77. A

78. D

If you missed two out of the previous three questions, begin your studying by turning to page 135, Powers and Exponents.

79. B

80. A

81. D

If you missed two out of the previous three questions, begin your studying by turning to page 138, Operations with Powers and Exponents or page 140, More Operations with Powers and Exponents.

82. E

83. A

84. D

If you missed two out of the previous three questions, begin your studying by turning to page 142, Square Roots and Cube Roots.

85. A

86. B

87. C

If you missed two out of the previous three questions, begin your studying by turning to page 149, Scientific Notation.

88. D

89. A

90. D

If you missed two out of the previous three questions, begin your studying by turning to page 163, Measurement of Basic Figures.

91. A

92. C

93. E

If you missed two out of the previous three questions, begin your studying by turning to page 179, Probability.

94. D

95. A

96. B

If you missed two out of the previous three questions, begin your studying by turning to page 189, Algebraic Expressions.

97. C

98. A

99. B

If you missed two out of the previous three questions, begin your studying by turning to page 194, Solving Simple Equations.

Chapter 1
The Basics

Collections of Numbers

It is important to feel comfortable with some terms, symbols, and operations before you review basic math and pre-algebra. Basic math involves many different collections of numbers. Understanding these will make it easier for you to understand basic math.

- ❑ **Natural or counting numbers:** 1, 2, 3, 4, 5, ...
- ❑ **Whole numbers:** 0, 1, 2, 3, 4, ...
- ❑ **Odd numbers:** Whole numbers not divisible by 2: 1, 3, 5, 7, 9, ...
- ❑ **Even numbers:** Whole numbers divisible by 2: 0, 2, 4, 6, 8, ...
- ❑ **Integers:** ... −2, −1, 0, 1, 2, ...
- ❑ **Negative integers:** ... −5, −4, −3, −2, −1
- ❑ **Rational numbers:** All fractions, such as: $-\frac{5}{6}$, $-\frac{7}{3}$, $\frac{3}{4}$, $\frac{9}{8}$. Every integer is a rational number (for example, the integer 2 can be written as $\frac{2}{1}$). All rational numbers can be written in the general form $\frac{a}{b}$, where a can be any integer and b can be any natural number. Rational numbers also include repeating decimals (such as 0.66...) and terminating decimals (such as 0.4), because they can be written in fraction form.
- ❑ **Irrational numbers:** Numbers that cannot be written as fractions, such as $\sqrt{2}$ and π (the Greek letter pi).

Ways to Show Multiplication and Division

We assume that you know the basics of multiplying and dividing single-digit numbers and have some idea of what these operations mean (multiplying 5 by 3 means if you had five groups of three things, you would have 15 things altogether; dividing 15 by 5 means if you had 15 things and divided them equally among 5 people, each person would get 3 things).

You can write these operations in many different ways, and it is important to recognize these variations. Some of the ways of indicating the multiplication of two numbers include the following:

- ❑ **Multiplication sign:** $2 \times 3 = 6$
- ❑ **Multiplication dot:** $2 \cdot 3 = 6$
- ❑ **An asterisk** (especially with computers and calculators): $2 * 3 = 6$

- ❏ **One set of parentheses:** $2(3) = 6$ or $(2)3 = 6$
- ❏ **Two sets of parentheses:** $(2)(3) = 6$
- ❏ **A variable** (letter) next to a number: $2a$ means 2 times a.
- ❏ **Two variables** (letters) next to each other: ab means a times b.

Some of the ways of indicating the division of two numbers include the following:

- ❏ **Division sign:** $6 \div 3 = 2$
- ❏ **Fraction bar:** $6/3 = 2$, $\frac{6}{3} = 2$, or $\frac{6}{3} = 2$

Multiplying and Dividing Using Zero

Any number multiplied by zero equals zero (because several groups each having zero items in them means a total of zero items).

$$0 \times 3 = 0$$
$$8 * 9 * 0 * 4 = 0$$
$$(0)(10) = 0$$
$$2a \cdot 0 = 0$$

Zero divided by any number except zero is zero (if you have nothing, dividing it among several people still leaves nothing for each person).

$$0 \div 5 = 0$$
$$0/3 = 0$$

Any number divided by zero is undefined (if you have several things, it doesn't make any sense to divide them among zero people).

$$4 \div 0 \text{ is undefined}$$
$$\frac{0}{0} \text{ is undefined}$$

Symbols and Terminology

The following are commonly used symbols in basic math and algebra. It is important to know what each symbol represents.

$=$ is equal to

$\neq$ is not equal to

$<$ is less than

$\not<$ is not less than

$>$ is greater than

$\not>$ is not greater than

$\leq$ is less than or equal to

$\not\leq$ is not less than or equal to

$\geq$ is greater than or equal to

$\not\geq$ is not greater than or equal to

$\approx$ is approximately equal to

Some Fundamental Properties

The four operations of addition, subtraction, multiplication, and division follow certain rules. Knowing these rules, and having names for them, is important before moving on.

Some Properties (Axioms) of Addition

❏ **Closure** is when all answers fall into the original set. When two even numbers are added, the answer will be an even number ($4 + 6 = 10$); thus, the set of even numbers has closure under addition. When two odd numbers are added, the answer is not an odd number ($1 + 3 = 4$); therefore, the set of odd numbers does not have closure under addition.

❏ The **commutative property of addition** means that the order of the numbers added does not matter.

$$3 + 4 = 4 + 3$$
$$7 + 10 = 10 + 7$$
$$a + b = b + a$$

Note: The commutative property does not hold true for subtraction.

$$5 - 3 \neq 3 - 5$$
$$a - b \neq b - a$$

❏ The **associative property of addition** means that the way the numbers are grouped does not matter.

$$(1 + 2) + 3 = 1 + (2 + 3)$$
$$(3 + 5) + 7 = 3 + (5 + 7)$$
$$a + (b + c) = (a + b) + c$$

The parentheses have moved, but both sides of the equation are still equal.

Note: The associative property does not hold true for subtraction.

$$(7 - 5) - 3 \neq 7 - (5 - 3)$$
$$(3 - 2) - 1 \neq 3 - (2 - 1)$$
$$(a - b) - c \neq a - (b - c)$$

❏ The **identity element** for addition is 0. When 0 is added to any number, it gives the original number.

$$5 + 0 = 5$$
$$0 + 3 = 3$$
$$a + 0 = a$$

❏ The **additive inverse** is the opposite or negative of the number. The sum of any number and its additive inverse is 0 (the identity).

$$2 + (-2) = 0;\ \text{thus, 2 and } -2 \text{ are additive inverses.}$$
$$\tfrac{5}{3} + {}^{-}\tfrac{5}{3} = 0;\ \text{thus, } \tfrac{5}{3} \text{ and } {}^{-}\tfrac{5}{3} \text{ are additive inverses.}$$
$$b + (-b) = 0;\ \text{thus, } b \text{ and } -b \text{ are additive inverses.}$$

Some Properties (Axioms) of Multiplication

❑ **Closure** is when all answers fall into the original set. When two even numbers are multiplied, they produce an even number ($2 \times 4 = 8$); thus, the set of even numbers has closure under multiplication. When two odd numbers are multiplied, the answer is an odd number ($3 \times 5 = 15$); thus, the set of odd numbers has closure under multiplication.

❑ **The commutative property of multiplication** means that the order of the numbers multiplied does not matter.

$$2 \times 3 = 3 \times 2$$
$$5(6) = (6)5$$
$$ab = ba$$

Note: The commutative property does not hold true for division.

$$6 \div 3 \neq 3 \div 6$$

❑ **The associative property of multiplication** means that the way the numbers are grouped does not matter.

$$(3 \times 4) \times 5 = 3 \times (4 \times 5)$$
$$(5 \times 7) \times 9 = 5 \times (7 \times 9)$$
$$(a \times b) \times c = a \times (b \times c)$$

The parentheses have moved, but both sides of the equation are still equal.

Note: The associative property does not hold true for division.

$$(12 \div 2) \div 3 \neq 12 \div (2 \div 3)$$

❑ The **identity element** for multiplication is 1. Any number multiplied by 1 gives the original number.

$$6 \times 1 = 6$$
$$9 \times 1 = 9$$
$$b \times 1 = b$$

❑ The **multiplicative inverse** is the **reciprocal** of the number, which means one divided by the number. When a number is multiplied by its reciprocal, the answer is 1.

$3 \times \frac{1}{3} = 1$; thus, 3 and $\frac{1}{3}$ are multiplicative inverses. $\frac{2}{3} \times \frac{3}{2} = 1$;

thus, $\frac{2}{3}$ and $\frac{3}{2}$ are multiplicative inverses.

$a \times \frac{1}{a} = 1$; thus, a and $\frac{1}{a}$ are multiplicative inverses, as long as $a \neq 0$.

A Property of Two Operations

The **distributive property** is when the number on the outside of the parentheses is distributed to each term on the inside of the parentheses. Multiplication distributes over addition or subtraction.

$$3(4 + 5) = 3(4) + 3(5)$$
$$6(4 - 2) = 6(4) - 6(2)$$
$$a(b + c) = a(b) + a(c)$$

Note: The distributive property cannot be used with other combinations of operations.

$$2(3 \times 4 \times 5) \neq 2(3) \times 2(4) \times 2(5)$$
$$7 + (2 - 1) \neq (7 + 2) - (7 + 1)$$

Grouping Symbols and Order of Operations

Numbers or variables commonly are grouped using three types of symbols: parentheses ( ), brackets [ ], and braces { }. Of these three, parentheses are used most often. Operations inside grouping symbols are the first to be worked out; they must be performed before all other operations.

Example Problems

These problems show the answers and solutions.

1. Simplify $2(3 + 4)$.

   answer: 14

   $2(3 + 4) = 2(7) = 14$

2. Simplify $4(5 - 3)$.

   answer: 8

   $4(5 - 3) = 4(2) = 8$

3. Simplify $(1 + 4)(2 + 3)$.

   answer: 25

   $(1 + 4)(2 + 3) = (5)(5) = 25$

4. Simplify $(9 - 7)(8 - 4)$.

   answer: 8

   $(9 - 7)(8 - 4) = (2)(4) = 8$

   Brackets and braces are used less often than parentheses. In the order of operations, parentheses should be used first, then brackets, and then braces: { [( )] }. Larger parentheses sometimes are used in place of brackets and braces.

5. Simplify $[(5 - 3) \times 7]$.

   answer: 14

   We work from the inside out:

   $$[(5 - 3) \times 7] = [(2) \times 7]$$
   $$= 14$$

6. Simplify $3\{4 + [2(1 + 3) + 5]\}$.

    answer: 51

    $$
    \begin{aligned}
    3\{4 + [2(1 + 3) + 5]\} &= 3\{4 + [2(4) + 5]\} \\
    &= 3\{4 + [8 + 5]\} \\
    &= 3\{4 + [13]\} \\
    &= 3\{17\} \\
    &= 51
    \end{aligned}
    $$

7. Simplify $2\left(4\left(2(5 - 3)\right) + 4\right)$.

    answer: 40

    $$
    \begin{aligned}
    2(4(2(5 - 3)) + 4) &= 2(4(2(2)) + 4) \\
    &= 2(4(4) + 4) \\
    &= 2(16 + 4) \\
    &= 2(20) \\
    &= 40
    \end{aligned}
    $$

8. Simplify $2\{10 - [3(1 + 4) - 6]\}$.

    answer: 2

    $$
    \begin{aligned}
    2\{10 - [3(1 + 4) - 6]\} &= 2\{10 - [3(5) - 6]\} \\
    &= 2\{10 - [15 - 6]\} \\
    &= 2\{10 - [9]\} \\
    &= 2\{1\} \\
    &= 2
    \end{aligned}
    $$

Order of Operations

The order of operations is important if multiplication, division, exponents, addition, subtraction, parentheses, and so on, are all in the same problem. The order in which these operations should be carried out is as follows

1. Parentheses
2. Exponents
3. Multiplication and division, from left to right
4. Addition and subtraction, from left to right

Example Problems

These problems show the answers and solutions.

1. Simplify $3 \times 4 + 2$.

    answer: 14

First do the multiplication:

$$3 \times 4 + 2 = 12 + 2$$

Then the addition:

$$12 + 2 = 14$$

2. Simplify $7 - 4 \div 2$.

    answer: 5

    Division comes first:

    $$7 - 4 \div 2 = 7 - 2$$

    Then the subtraction:

    $$7 - 2 = 5$$

3. Simplify $12 - 4 + 2 \times 10^2 + (5 + 3) - 3$.

    answer: 213

    First do the operation inside the parentheses:

    $$12 - 4 + 2 \times 10^2 + (5 + 3) - 3 = 12 - 4 + 2 \times 10^2 + (8) - 3$$

    Then apply the exponent:

    $$12 - 4 + 2 \times 10^2 + 8 - 3 = 12 - 4 + 2 \times 100 + 8 - 3$$

    Next multiply:

    $$12 - 4 + 2 \times 100 + 8 - 3 = 12 - 4 + 200 + 8 - 3$$

    Finally, carry out the addition and subtraction starting on the left:

    $$\begin{aligned} 12 - 4 + 200 + 8 - 3 &= 8 + 200 + 8 - 3 \\ &= 208 + 8 - 3 \\ &= 216 - 3 \\ &= 213 \end{aligned}$$

4. Simplify $100 - 2[1 + 5(3 + 2^2)]$.

    answer: 28

First do the work inside the parentheses, but because there is an exponent inside the parentheses, you must apply the exponent before adding what is inside the parentheses:

$$100 - 2[1 + 5(3 + 2^2)] = 100 - 2[1 + 5(3 + 4)]$$
$$= 100 - 2[1 + 5(7)]$$

Then do the multiplication inside the brackets:

$$100 - 2[1 + 5(7)] = 100 - 2[1 + 35]$$

Next finish the operation inside the brackets:

$$100 - 2[1 + 35] = 100 - 2[36]$$

Then do the multiplication:

$$= 100 - 72$$

And, finally, finish with the subtraction:

$$100 - 72 = 28$$

Note: A handy phrase some people like for remembering the order of operations is: **P**lease **E**xcuse **M**y **D**ear **A**unt **S**ally, where the first letters are reminders to work in the order **P**arentheses, **E**xponents, **M**ultiplication/**D**ivision, **A**ddition/**S**ubtraction).

Chapter 2
Whole Numbers

Place Value

Our modern number system is a **place value system**. Each place receives a specific value. For instance, in the number 856, the 8 is in the hundreds place, the 5 is in the tens place, and the 6 is in the ones place. The number system is designed around powers of 10, such that $10^0 = 1$, $10^1 = 10$, $10^2 = 10 \times 10 = 100$, $10^3 = 10 \times 10 \times 10 = 1000$, and so on. These progressive powers of 10 are shown in Table 2-1.

Table 2-1						
Millions	**Hundred Thousands**	**Ten Thousands**	**Thousands**	**Hundreds**	**Tens**	**Ones**
1,000,000	100,000	10,000	1,000	100	10	1
10^6	10^5	10^4	10^3	10^2	10^1	10^0
				8	5	6

Notice how 856 fits into the grid.

The place value of each digit can also be shown by writing the numbers in **expanded notation**.

Example Problems

These problems show the answers and solutions.

1. Write 354 in expanded notation.

   answer: $(3 \times 10^2) + (5 \times 10^1) + (4 \times 10^0)$

$$354 = 300 + 50 + 4$$
$$= (3 \times 100) + (5 \times 10) + (4 \times 1)$$

2. Write 35,709 in expanded notation.

   answer: $(3 \times 10^4) + (5 \times 10^3) + (7 \times 10^2) + (0 \times 10^1) + (9 \times 10^0)$

$$35,709 = 30,000 + 5,000 + 700 + 9$$
$$= (3 \times 10,000) + (5 \times 1,000) + (7 \times 100) + (0 \times 10) + (9 \times 1)$$

You also may be given the number in expanded notation and asked to simplify the number.

3. Write $(4 \times 10^2) + (2 \times 10^1) + (5 \times 10^0)$ in simplest form.

    answer: 425

$$(4 \times 10^2) + (2 \times 10^1) + (5 \times 10^0) = (4 \times 100) + (2 \times 10) + (5 \times 1)$$
$$= 400 + 20 + 5$$

4. Write $(2 \times 10^3) + (6 \times 10^2) + (3 \times 10^1) + (8 \times 10^0)$ in simplest form.

    answer: 2638

$$(2 \times 10^3) + (6 \times 10^2) + (3 \times 10^1) + (8 \times 10^0) = (2 \times 1000) + (6 \times 100) + (3 \times 10) + (8 \times 1)$$
$$= 2000 + 600 + 30 + 8$$

Work Problems

Use these problems to give yourself additional practice.

1. Write 857 in expanded notation.

2. Write 3007 in expanded notation.

3. Write $(8 \times 10^3) + (3 \times 10^2) + (1 \times 10^1) + (9 \times 10^0)$ in simplest form.

Worked Solutions

1. $(8 \times 10^2) + (5 \times 10^1) + (7 \times 10^0)$

$$857 = 800 + 50 + 7$$
$$= (8 \times 100) + (5 \times 10) + (7 \times 1)$$
$$= (8 \times 10^2) + (5 \times 10^1) + (7 \times 10^0)$$

2. $(3 \times 10^3) + (0 \times 10^2) + (0 \times 10^1) + (7 \times 10^0)$

$$3007 = 3000 + 7$$
$$= (3 \times 1000) + (0 \times 100) + (0 \times 10) + (7 \times 1)$$
$$= (3 \times 10^3) + (0 \times 10^2) + (0 \times 10^1) + (7 \times 10^0)$$

3. **8319**

$$(8 \times 10^3) + (3 \times 10^2) + (1 \times 10^1) + (9 \times 10^0) = (8 \times 1000) + (3 \times 100) + (1 \times 10) + (9 \times 1)$$
$$= 8000 + 300 + 10 + 9$$
$$= 8319$$

Computation

Some people find addition, subtraction, multiplication, and division of whole numbers to be difficult, but if you learn them step-by-step and practice, you'll discover that they aren't so hard after all.

Adding and Subtracting

To add or subtract whole numbers, line up the numbers from the last digit, the ones place, and work from right to left.

Example Problems

These problems show the answers and solutions.

1. Add 34 + 13 + 21.

    answer: 68

    $$
    \begin{array}{r}
    34 \\
    13 \\
    +21 \\
    \hline
    68
    \end{array}
    $$

2. Add 19 + 17 + 20.

    answer: 56

    When the sum of the digits in any single column yields a two-digit number, you must "carry" the tens digit of that result and place it above the next rightmost column of numbers:

    $$
    \begin{array}{r}
    \overset{1}{1}9 \\
    17 \\
    +\,20 \\
    \hline
    56
    \end{array}
    $$

3. Add 8423 + 279.

    answer: 8702

    $$
    \begin{array}{r}
    \overset{1\,1}{8}423 \\
    +\,279 \\
    \hline
    8702
    \end{array}
    $$

4. Subtract 59 − 13.

    answer: 46

    $$\begin{array}{r} 59 \\ -13 \\ \hline 46 \end{array}$$

5. Subtract 60 − 35.

    answer: 25

    When the top digit in a column isn't large enough to subtract the bottom digit from it (as in this case, where you can't take 5 away from 0), you must "borrow" from the next rightmost column, decreasing it by 1 and increasing the column you're working with by 10:

    $$\begin{array}{r} \overset{5}{\cancel{6}}{}^{1}0 \\ -35 \\ \hline 25 \end{array}$$

6. Subtract 52 − 15.

    answer: 37

    $$\begin{array}{r} \overset{4}{\cancel{5}}{}^{1}2 \\ -15 \\ \hline 37 \end{array}$$

7. Subtract 345 − 217.

    answer: 128

    $$\begin{array}{r} 34\overset{3}{\cancel{}}{}^{1}5 \\ -217 \\ \hline 128 \end{array}$$

Work Problems

Use these problems to give yourself additional practice.

1. Add 87 + 25 + 11.

2. Add 758 + 2153.

3. Subtract 94 − 34.

4. Subtract 937 − 183.

5. Subtract 31,535 − 543.

Worked Solutions

1. **123**

    $$\begin{array}{r} \overset{11}{87} \\ 25 \\ +11 \\ \hline 123 \end{array}$$

2. **2911**

    $$\begin{array}{r} \overset{11}{758} \\ +2153 \\ \hline 2911 \end{array}$$

3. **60**

    $$\begin{array}{r} 94 \\ -34 \\ \hline 60 \end{array}$$

4. **754**

    $$\begin{array}{r} \overset{8}{9}37 \\ -183 \\ \hline 754 \end{array}$$

5. **30992**

    $$\begin{array}{r} 31\overset{0\ 14}{5}35 \\ -543 \\ \hline 30992 \end{array}$$

Multiplying

Complete multiplication problems from right to left and use basic multiplication. You need to know your multiplication table up to 9×9 in order to perform these procedures.

Example Problems

These problems show the answers and solutions.

1. Multiply 6×4.

    answer: 24

    $$\begin{array}{r} 6 \\ \times 4 \\ \hline 24 \end{array}$$

2. Multiply 22×5.

    answer: 110

    $$\begin{array}{r} \overset{1}{22} \\ \times 5 \\ \hline 110 \end{array}$$

3. Multiply 12×12.

    answer: 144

    When there's more than one digit in the second number, first multiply by its ones digit, then by its tens digit—writing this result starting one column to the left—and so on; then add up these subtotals:

$$
\begin{array}{r}
12 \\
\times 12 \\
\hline
24 \\
+12 \\
\hline
144
\end{array}
$$

4. Multiply 432×200.

    answer: 86,400

$$
\begin{array}{r}
432 \\
\times 200 \\
\hline
000 \\
000 \\
+864 \\
\hline
86400
\end{array}
$$

Work Problems

Use these problems to give yourself additional practice.

1. Multiply 23×8.

2. Multiply 73×12.

3. Multiply 5702×62.

4. Multiply 684×73.

5. Multiply 187×307.

Worked Solutions

1. **184**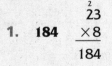

2. **876**
$$
\begin{array}{r}
73 \\
\times 12 \\
\hline
146 \\
+73 \\
\hline
876
\end{array}
$$

3. **353,524**
$$
\begin{array}{r}
5702 \\
\times 62 \\
\hline
11404 \\
+34212 \\
\hline
353524
\end{array}
$$

4. **49,932**
$$
\begin{array}{r}
684 \\
\times 73 \\
\hline
2052 \\
+4788 \\
\hline
49932
\end{array}
$$

5. **57,409**
$$
\begin{array}{r}
187 \\
\times 307 \\
\hline
1309 \\
000 \\
+561 \\
\hline
57409
\end{array}
$$

Dividing

Think of division as finding out how many groups a number can be separated into, or the amount contained within each group. For example: $20 \div 4 = 5$. Assume that you have 20 pencils and you want to split them up into 4 groups of pencils. The number of pencils that you then have in each of these groups will be 5.

Each different number involved in this has a name, and although the names can be a little confusing, we'll use them here so we can talk about which is which in the problems that follow. The 20, the number that is being divided into parts, is called the **dividend**. The 4, the number of groups into which the pencils are being divided, is called the **divisor**. Finally the 5, our answer, is called the **quotient**.

Example Problems

These problems show the answers and solutions.

1. Divide $10 \div 2$.

   answer: 5

Begin by writing this: $2\overline{)10}$

Notice that the dividend, 10, is written inside the long division sign, while the divisor, 2, is written outside. This is opposite the arrangement in the original problem, but it's important to how the procedure works.

Then ask yourself "What's the biggest number I could multiply 2 by without getting more than 10?" and, maybe after some trial-and-error or scratch work, decide on 5. Write the 5 above the long division sign like this:

$$\begin{array}{r} 5 \\ 2\overline{)10} \end{array}$$

Next take the 5 and multiply it times the divisor, 2, which yields 10, and write this below our dividend 10 like this:

$$\begin{array}{r} 5 \\ 2\overline{)10} \\ 10 \end{array}$$

Finally subtract down, like this:

$$\begin{array}{r} 5 \\ 2\overline{)10} \\ -10 \\ \hline 0 \end{array}$$

The fact that you got a 0 at the end of this process means that 10 things can be divided evenly into 2 groups—there aren't any leftovers. Most of the time in this chapter we'll stick to division problems that come out evenly, although in the following example we'll see what can happen. We'll deal more with the possibility of division problems that don't come out evenly in later chapters.

2. Divide $51 \div 2$.

    answer: 25, with a remainder of 1.

    The procedure is just as before, but in this case there are more digits. Start with the leftmost digit:

$$\begin{array}{r} 2 \\ 2\overline{)51} \\ -4 \\ \hline 11 \end{array}$$

Notice that after subtracting the 4 from the 5, you drop the next digit of the dividend (the 1) down beside that result. Now ask yourself what is the largest number that could be multiplied by 2 without getting a result bigger than 11. Because the answer to this question is 5, continue:

$$
\begin{array}{r}
25 \\
2\overline{)\,51} \\
-4 \\
\hline
11 \\
-10 \\
\hline
1
\end{array}
$$

This time you were left with a 1 at the end of the procedure. This number is called a **remainder**, and it means that when you have 51 objects and you split them into two groups, it's not possible to put equal whole numbers in each group—there will be one item left over. We'll talk more in later chapters about different ways of handling remainders.

3. Divide 225 ÷ 15.

   answer: 15

$$
\begin{array}{r}
15 \\
15\overline{)\,225} \\
-15 \\
\hline
75 \\
-75 \\
\hline
0
\end{array}
$$

4. Divide 2016 ÷ 36.

   answer: 56

$$
\begin{array}{r}
56 \\
36\overline{)\,2016} \\
-180 \\
\hline
216 \\
-216 \\
\hline
0
\end{array}
$$

Work Problems
Use these problems to give yourself additional practice.

1. Divide 72 ÷ 3.

2. Divide 285 ÷ 5.

3. Divide 3372 ÷ 12.

4. Divide 876 ÷ 73.

5. Divide 3293 ÷ 37.

Worked Solutions

$$
\begin{array}{r}
24 \\
3\overline{)\,72} \\
-6 \\
\hline
12 \\
-12 \\
\hline
0
\end{array}
$$

1. **24**

$$
\begin{array}{r}
57 \\
5\overline{)\,285} \\
-25 \\
\hline
35 \\
-35 \\
\hline
0
\end{array}
$$

2. **57**

$$
\begin{array}{r}
281 \\
12\overline{)\,3372} \\
-24 \\
\hline
97 \\
-96 \\
\hline
12 \\
-12 \\
\hline
0
\end{array}
$$

3. **281**

$$
\begin{array}{r}
12 \\
73\overline{)\,876} \\
-73 \\
\hline
146 \\
-146 \\
\hline
0
\end{array}
$$

4. **12**

(Notice how this problem compares with work problem 2 in the "Multiplying" section earlier!)

$$
\begin{array}{r}
89 \\
37\overline{)\,3293} \\
-296 \\
\hline
333 \\
-333 \\
\hline
0
\end{array}
$$

5. **89**

Rounding Off

To **round off** any number,

1. Underline the value of the digit to which you are rounding off.

2. Look one place to the right of the underlined digit.

3. If this digit is a 5 or higher, increase the underlined value by 1 and change all the numbers to the right of this value to zeros. If the digit is less than 5, leave the underlined digit as it is and change all the digits to its right to zero.

Example Problems

These problems show the answers and solutions.

1. Round 354 to the nearest ten.

   answer: 3_5_4 is rounded down to 350.

2. Round 64,278 to the nearest hundred.

   answer: 64,_2_78 is rounded up to 64,300.

3. Round 725,369 to the nearest ten thousand.

   answer: 7_2_5,369 is rounded up to 730,000.

Work Problems

Use these problems to give yourself additional practice.

1. Round 763 to the nearest ten.

2. Round 763 to the nearest hundred.

3. Round 459,732 to the nearest thousand.

Worked Solutions

1. **760**   7_6_3 is rounded down to 760.

2. **800**   _7_63 is rounded up to 800.

3. **460,000**   459,_7_32 is rounded up to 460,000. (Notice how carrying affects this!)

Estimating

Being able to estimate can save you time and is useful when checking your answers to see whether your solution is reasonable.

Estimating Sums and Differences

Rounded numbers can be used to estimate sums and differences.

Example Problems

These problems show the answers and solutions.

1. Estimate the sum of 465 + 779 by rounding to the nearest ten.

   answer: 1250

   $465 + 779 \approx 470 + 780 = 1250$

2. Estimate the sum of 53,642 + 21,799 by rounding to the nearest thousand.

   answer: 76,000

   $53,642 + 21,799 \approx 54,000 + 22,000 = 76,000$

3. Estimate the difference of 578 − 231 by rounding to the nearest ten.

   answer: 350

   $578 - 231 \approx 580 - 230 = 350$

4. Estimate the difference of 547,925 − 239,548 by rounding to the nearest ten thousand.

   answer: 310,000

   $547,925 - 239,548 \approx 550,000 - 240,000 = 310,000$

Work Problems

Use these problems to give yourself additional practice.

1. Estimate the sum of 89 + 123 by rounding to the nearest ten.

2. Estimate the sum of 8742 + 13,169 by rounding to the nearest thousand.

3. Estimate the difference of 764 − 349 by rounding to the nearest hundred.

Worked Solutions

1. **210** $89 + 123 \approx 90 + 120 = 210$

2. **22,000** $8742 + 13,169 \approx 9000 + 13,000 = 22,000$

3. **500** $764 - 349 \approx 800 - 300 = 500$

Notice that in this last problem, we're not very close to the exact answer, which would be 415; but this was much easier than working out the exact answer, so we have a tradeoff between ease and correctness.

Estimating Products and Quotients

Rounded numbers can be used to estimate products and quotients.

Example Problems

These problems show the answers and solutions.

1. Estimate the product of 24×36 by rounding to the nearest ten.

    answer: 800

    $24 \times 36 \approx 20 \times 40 = 800$

2. Estimate the product of 627×194 by rounding to the nearest hundred.

    answer: 120,000

    $627 \times 194 \approx 600 \times 200 = 120,000$

    When the digit you're rounding at is close to 5, your approximation can be poor. If both numbers are close to 5, a more accurate product can be found by rounding one number up and the other number down.

3. Estimate the product of 450×250 by rounding to the nearest hundred.

    answer: 120,000

    $450 \times 250 \approx 400 \times 300 = 120,000$ (One number is rounded up, while the other is rounded down.)

    This can also be done by rounding the first number up and the second number down.

    $450 \times 250 \approx 500 \times 200 = 100,000$

    Rounding one number up and one down gives a more accurate approximation than rounding both numbers up.

4. Estimate the quotient of $63 \div 28$ by rounding to the nearest ten.

    answer: 2

    $63 \div 28 \approx 60 \div 30 = 2$

5. Estimate the quotient of 789 ÷ 175 by rounding to the nearest hundred.

    answer: 4

    789 ÷ 175 ≈ 800 ÷ 200 = 4

6. Estimate the quotient of 15,103 ÷ 4699 by rounding to the nearest thousand.

    answer: 3

    15,103 ÷ 4699 ≈ 15,000 ÷ 5000 = 3

7. Estimate the quotient of 250,423 ÷ 5354 by rounding to the nearest thousand.

    answer: 50

    250,423 ÷ 5354 ≈ 250,000 ÷ 5000 = 50

Work Problems
Use these problems to give yourself additional practice.

1. Estimate the product of 42 × 41 by rounding to the nearest ten.

2. Estimate the product of 617 × 68 by rounding to the nearest ten.

3. Estimate the quotient of 4346 ÷ 48 by rounding to the nearest ten.

4. Estimate the quotient of 16,044 ÷ 191 by rounding to the nearest hundred.

Worked Solutions

1. **1600** 42 × 41 ≈ 40 × 40 = 1600

2. **43,400** 617 × 68 ≈ 620 × 70 = 43,400

3. **87** 4346 ÷ 48 ≈ 4350 ÷ 50 = 87

4. **80** 16,044 ÷ 191 ≈ 16,000 ÷ 200 = 80

Divisibility Rules

If you ever need to know whether one number divides evenly into another, divisibility rules can help save time. Instead of working through long division, try using the rules in Table 2-2.

Table 2-2: Divisibility Rules	
A number is divisible by:	*when:*
2	it ends in 0, 2, 4, 6, or 8 (an even number).
3	the sum of its digits is divisible by 3.
4	the number created by the last two digits is divisible by 4.
5	it ends in 0 or 5.
6	it is divisible by both 2 and 3.
7	(no easy rule).
8	the number created by the last three digits is divisible by 8.
9	the sum of its digits is divisible by 9.

Example Problems

These problems show the answers and solutions.

1. Is 36 divisible by 2?

   answer: Yes.   36 ends in 6, making it divisible by 2.

2. Is 912 divisible by 3?

   answer: Yes.   The sum of the digits of 912 is 12, which is divisible by 3.

3. Is 126 divisible by 4?

   answer: No.   The number created by the last two digits is 26, which is not divisible by 4.

4. Is 145 divisible by 5?

   answer: Yes.   The last digit is 5, making it divisible by 5.

5. Is 1549 divisible by 6?

   answer: No.   1,549 is not divisible by 2, so it is not divisible by 6 either.

6. Is 3248 divisible by 8?

   answer: Yes.   The number created by the last three digits is 248, which is divisible by 8.

7. 350 is divisible by which of the following numbers: 2, 4, 5, 9?

   answer: 2, 5.   350 is divisible by 2 because the number ends in 0. 350 is divisible by 5 because the number ends in 0.

8. 378 is divisible by which of the following numbers: 2, 3, 4, 6, 8, 9?

    answer: 2, 3, 6, 9.　　378 is divisible by 2 because the number ends in 8, and it is divisible by 3 because the sum of its digits add up to 18, which is divisible by 3. The number is divisible by 6 because it is divisible by 2 and 3. It is also divisible by 9 because the sum of the digits is 18, which is divisible by 9.

Work Problems

Use these problems to give yourself additional practice.

1. Is 5671 divisible by 3?

2. Is 32,545 divisible by 5?

3. Is 1,243,342 divisible by 4?

4. Is 123,456,789 divisible by 9?

5. Is 43,522 divisible by 8?

Worked Solutions

1. **No.**　The sum of the digits of 5671 is 19, which is not divisible by 3.

2. **Yes.**　The last digit of 32,545 is 5, so it is divisible by 5.

3. **No.**　The number formed by the last two digits, 42, is not divisible by 4.

4. **Yes.**　The sum of the digits in 123,456,789 is 45, which is divisible by 9.

5. **No.**　The number formed by last three digits of 43,522 is not divisible by 8.

Factoring

The **factors** of a number are numbers that can be multiplied together to get that number. They are also sometimes called **divisors** because they're the numbers that will divide evenly into that number.

Example Problems

These questions show the answers and solutions.

1. What are the factors of 20?

    answer: 1, 2, 4, 5, 10, and 20

    $1 \times 20 = 20$

    $2 \times 10 = 20$

    $4 \times 5 = 20$

2.   What are the factors of 36?

answer: 1, 2, 3, 4, 6, 9, 12, 18, and 36

$1 \times 36 = 36$

$2 \times 18 = 36$

$3 \times 12 = 36$

$4 \times 9 = 36$

$6 \times 6 = 36$

3.   What are the factors of 100?

answer: 1, 2, 4, 5, 10, 20, 25, 50, and 100

$1 \times 100 = 100$

$2 \times 50 = 100$

$4 \times 25 = 100$

$5 \times 20 = 100$

$10 \times 10 = 100$

Work Problems

Use these problems to give yourself additional practice.

1.   What are the factors of 15?

2.   What are the factors of 72?

3.   What are the factors of 121?

Worked Solutions

1.   **1, 3, 5, and 15**

$$1 \times 15 = 15$$
$$3 \times 5 = 15$$

2.   **1, 2, 3, 4, 6, 8, 9, 12, 18, 24, 36, and 72**

$$1 \times 72 = 72$$
$$2 \times 36 = 72$$
$$3 \times 24 = 72$$

$$4 \times 18 = 72$$
$$6 \times 12 = 72$$
$$8 \times 9 = 72$$

3.   **1, 11, and 121**

$$1 \times 121 = 121$$
$$11 \times 11 = 121$$

Prime Numbers

A number (greater than 1) that can be divided evenly only by itself and 1 is called a **prime number**. A prime number has only two factors: itself and 1.

Example Problems

These problems show the answers and solutions.

1.   Is 5 a prime number?

     answer: Yes.   5 can only be divided evenly by 1 and 5.

2.   Is 16 a prime number?

     answer: No.   16 is divisible by other numbers. The factors of 16 are 1, 2, 4, 8, and 16, so it is not a prime number.

3.   Is 23 a prime number?

     answer: Yes.   23 is only divisible evenly by 1 and 23, making it a prime number.

Note: 2 is the only even number that is prime; after that all even numbers can be divided by 2, but the only factors of 2 are 1 and 2. Prime numbers do not include 0 and 1. There are 15 prime numbers less than 50. These include: 2, 3, 5, 7, 11, 13, 17, 19, 23, 29, 31, 37, 41, 43, and 47.

Work Problems

Use these problems to give yourself additional practice.

1.   Is 9 a prime number?

2.   Is 11 a prime number?

3.   Is 63 a prime number?

Worked Solutions

1.   **No.** $9 = 3 \times 3$, so 9 has factors other than 1 and itself.

2. **Yes.**   11 can be divided evenly only by 1 and itself.

3. **No.** $63 = 9 \times 7$, so 63 has divisors other than 1 and itself.

Composite Numbers

A number that is divisible by more than itself and 1 is called a **composite number**. Another way to define composite numbers is that they are positive numbers that have more than two factors. All even numbers, except 2, are composite numbers (4, 6, 8, 10,...). The numbers 0 and 1 are neither composite numbers nor prime numbers.

Example Problems

These problems show the answers and solutions.

1. Is 15 a composite number?

    answer: Yes.   15 is divisible by 3 and 5, making it a composite number. The factors of 15 are 1, 3, 5, and 15.

2. Is 19 a composite number?

    answer: No.   19 is a prime number; it can only be divided evenly by 19 and 1.

3. Is 99 a composite number?

    answer: Yes.   99 is divisible by 3, 9, 11, and 33, making it a composite number. The factors of 99 are 1, 3, 9, 11, 33, and 99.

Work Problems

Use these problems to give yourself additional practice.

1. Is 27 a composite number?

2. Is 31 a composite number?

3. Is 121 a composite number?

Worked Solutions

1. **Yes.** $27 = 3 \times 9$, so 27 is a composite number.

2. **No.**   31 is a prime number because its only divisors are 1 and itself.

3. **Yes.** $121 = 11 \times 11$, so 121 is a composite number.

Factor Trees

All composite numbers can be written as the product of prime numbers. Prime factors can be determined by using a factor tree. A factor tree looks like this:

$$16$$
$$2 \times 8$$
$$2 \times 2 \times 4$$
$$2 \times 2 \times 2 \times 2$$

This tree can also be created in another way:

$$16$$
$$4 \times 4$$
$$2 \times 2 \times 2 \times 2$$

Both ways are correct and show that no matter how 16 is factored, the prime factors will be the same (although in some cases they might be written in a different order).

Example Problems

These problems show the answers and solutions.

1. Using a factor tree, express 30 as a product of prime factors.

    answer:

    $$30$$
    $$2 \times 15$$
    $$2 \times 3 \times 5$$

    So $2 \times 3 \times 5$ is called the **prime factorization** of 30.

2. Using a factor tree, express 100 as a product of prime factors.

    answer:

    $$100$$
    $$2 \times 50$$
    $$2 \times 2 \times 25$$
    $$2 \times 2 \times 5 \times 5$$

So the prime factorization of 100 is $2 \times 2 \times 5 \times 5$, which can also be written as $2^2 \times 5^2$.

3.   Using a factor tree, express 54 as a product of prime factors.

answer:

$$54$$
$$6 \times 9$$
$$2 \times 3 \times 3 \times 3$$

This can also be written as 2×3^3.

Work Problems

Use these problems to give yourself additional practice.

1.   Using a factor tree, express 35 as a product of prime factors.

2.   Using a factor tree, express 44 as a product of prime factors.

3.   Using a factor tree, express 120 as a product of prime factors.

Worked Solutions

1.   **5×7**

2.   **$2 \times 2 \times 11$**

$$4 \times 11$$
$$2 \times 2 \times 11$$

This can also be written as $2^2 \times 11$.

3.

$$12 \times 10$$
$$6 \times 2 \times 5 \times 2$$
$$3 \times 2 \times 2 \times 5 \times 2$$

This can also be written as $2^3 \times 3 \times 5$.

Chapter 3
Fractions

Fractions are numbers that represent parts of things, like half of an apple or three quarters of an hour. We write $\frac{1}{2}$ when we mean one part of something that has been divided into two equal parts, or $\frac{3}{4}$ when we mean three parts of something that has been divided into four equal parts. The number above the bar is called the **numerator**, and the number below the bar is called the **denominator** (remembering that "denominator" starts with the same letter as "down" helps many people keep these straight). Fractions also are called **rational numbers** because they can be formed by taking the ratio of two integers, dividing one integer by another, as long as the second number isn't zero.

Not all fractions are less than one, and not all fractions are positive. A fraction like $-\frac{4}{5}$ might represent a height four fifths of a foot below normal. Negative fractions will be handled more fully in Chapter 6. A fraction like $\frac{3}{2}$ is a perfectly good way of representing something like a distance that is three half-miles. Fractions with numerators larger than their denominators are sometimes called **improper fractions**, and by contrast, fractions between zero and one are sometimes called **proper fractions**.

Also remember that any integer can be thought of as a fraction, for instance by writing 6 as $\frac{6}{1}$.

Equivalent Fractions

Fractions allow you many ways of writing the same amount. Half of an hour is the same amount of time as thirty minutes, or thirty sixtieths of an hour, so $\frac{1}{2} = \frac{30}{60}$. Knowing how to convert between equivalent fractions will be necessary for the other operations you will perform on fractions.

Example Problems
These problems show the answers and solutions.

1. Determine whether the fractions $\frac{2}{3}$ and $\frac{40}{60}$ are equivalent.

    answer: The fractions are equivalent.

    You multiply the denominator of each fraction by the numerator of the other. If the results are equal, then the fractions are equivalent:

    $$\frac{2}{3} \times \frac{40}{60}$$
    $$120 = 120$$

2. Determine whether the fractions $\frac{5}{6}$ and $\frac{9}{12}$ are equivalent.

    answer: The fractions are not equivalent.

    $$\frac{5}{6} \times \frac{9}{12}$$
    $$60 \neq 54$$

Reducing Fractions

A fraction like $\frac{1}{2}$ is said to be in **reduced** form (or simplest form) because there is no way to write it as an equivalent fraction with a smaller denominator.

Example Problems

These problems show the answers and solutions.

1. Write the fraction $\frac{4}{6}$ in reduced form.

    answer: $\frac{2}{3}$

    First write both the numerator and denominator in factored form. The key is to find all common factors in the numerator and denominator and cancel them:

    $$\frac{4}{6} = \frac{2 \cdot 2}{2 \cdot 3} = \frac{2 \cdot \not{2}}{\not{2} \cdot 3} = \frac{2}{3}$$

    The reason this procedure works is that the 2s you cancelled out offset each other. A more complete way of writing the solution looks like this:

    $$\frac{4}{6} = \frac{2 \cdot 2}{2 \cdot 3} = \frac{2}{3} \cdot \frac{2}{2} = \frac{2}{3} \cdot 1 = \frac{2}{3}$$

    This depends on recognizing that a fraction with the same numerator and denominator, like $\frac{2}{2}$, is just 1, because cutting something in half but then keeping both halves preserves the amount you started with. This complete version of the computation isn't often written out, but it explains why the procedure works.

2. Write the fraction $\frac{12}{60}$ in reduced form.

    answer: $\frac{1}{5}$

    $$\frac{12}{60} = \frac{2 \cdot 2 \cdot 3}{2 \cdot 2 \cdot 3 \cdot 5} = \frac{\not{2} \cdot \not{2} \cdot \not{3}}{\not{2} \cdot \not{2} \cdot \not{3}} \cdot \frac{1}{5} = \frac{1}{5}$$

Notice that you write a 1 on top of the fraction, even though everything else cancels.

3. Write the fraction $\frac{120}{100}$ in reduced form.

    answer: $\frac{6}{5}$

$$\frac{120}{100} = \frac{2 \cdot \cancel{2} \cdot \cancel{2} \cdot 3 \cdot \cancel{5}}{\cancel{2} \cdot \cancel{2} \cdot 5 \cdot \cancel{5}} = \frac{2 \cdot 3}{5} = \frac{6}{5}$$

Enlarging Denominators

It is also sometimes necessary to write a fraction as an equivalent fraction with a larger denominator.

Example Problems

These problems show answers and solutions.

1. Write the fraction $\frac{2}{3}$ as an equivalent fraction with a denominator of 6.

    answer: $\frac{4}{6}$

    Use the opposite of the procedure for reducing fractions. Because the denominator must be multiplied by 2 to produce 6, multiply both the top and bottom of the fraction by 2 (which is allowed for the same reason canceling was). It looks like this:

$$\frac{2}{3} = \frac{2 \cdot 2}{3 \cdot 2} = \frac{4}{6}$$

    Notice how this compares to what you did in Example Problem 1 under "Reducing Fractions."

2. Write $\frac{1}{4}$ as an equivalent fraction with a denominator of 24.

    answer: $\frac{6}{24}$

    Because the current denominator of 4 needs to be multiplied by 6 to make 24, proceed as follows:

$$\frac{1}{4} = \frac{1 \cdot 6}{4 \cdot 6} = \frac{6}{24}$$

3. Write $\frac{1}{5}$ as an equivalent fraction with a denominator of 100.

    answer: $\frac{20}{100}$

$$\frac{1}{5} = \frac{1 \cdot 20}{5 \cdot 20} = \frac{20}{100}$$

Work Problems

Use these problems to give yourself additional practice.

1. Write the fraction $\frac{9}{12}$ in reduced form.

2. Write the fraction $\frac{28}{24}$ in reduced form.

3. Write the fraction $\frac{25}{100}$ in reduced form.

4. Write $\frac{1}{4}$ as an equivalent fraction with a denominator of 12.

5. Write $\frac{7}{12}$ as an equivalent fraction with a denominator of 24.

Worked Solutions

1. $\frac{3}{4}$ $\quad \frac{9}{12} = \frac{3 \cdot \cancel{3}}{2 \cdot 2 \cdot \cancel{3}} = \frac{3}{4}$

2. $\frac{7}{6}$ $\quad \frac{28}{24} = \frac{\cancel{2} \cdot \cancel{2} \cdot 7}{\cancel{2} \cdot \cancel{2} \cdot 2 \cdot 3} = \frac{7}{6}$

3. $\frac{1}{4}$ $\quad \frac{25}{100} = \frac{\cancel{5} \cdot \cancel{5}}{2 \cdot 2 \cdot \cancel{5} \cdot \cancel{5}} = \frac{1}{4}$

4. $\frac{3}{12}$ $\quad \frac{1}{4} = \frac{1 \cdot 3}{4 \cdot 3} = \frac{3}{12}$

5. $\frac{14}{24}$ $\quad \frac{7}{12} = \frac{7 \cdot 2}{12 \cdot 2} = \frac{14}{24}$

Mixed Numbers

It is traditional sometimes to write fractions that are bigger than one (this means their numerator is bigger than their denominator) as **mixed numbers**. A mixed number is an integer followed by a proper fraction. $\frac{3}{2}$ miles, for instance, is the same as one mile and another half mile, or $1\frac{1}{2}$ miles. One advantage of mixed numbers is that it's easier to tell immediately how big they are compared to each other: $10\frac{1}{4}$ is smaller than $11\frac{1}{3}$, but that's not as easy to tell when they're written as improper fractions $\frac{41}{4}$ and $\frac{34}{3}$. On the other hand, operations like addition, subtraction, multiplication, and division will often be easier when you use improper fractions. So sometimes one form is better, and other times it's not, and it's important to be able to switch back and forth.

Example Problems

These problems show the answers and solutions.

1. Convert the improper fraction $\frac{6}{5}$ to a mixed number.

    answer: $1\frac{1}{5}$

Use division, from the previous chapter:

$$5\overline{)\,6}$$
$$\underline{-5}$$
$$1$$

Notice that, unlike most of the examples in the previous chapter, this time you didn't end up with a 0 at the end. The remainder of 1 means that when you divide, say, 6 sandwiches among 5 hungry people, you'll first give each person 1 sandwich and have 1 sandwich left over. If you divide this remaining sandwich into 5 equal parts, then each person could have 1 sandwich plus $\frac{1}{5}$ of the leftover.

2.   Convert $\frac{33}{6}$ to a mixed number.

   answer: $5\frac{3}{6}$, which you can also reduce to $5\frac{1}{2}$.

$$6\overline{)\,33}$$
$$\underline{-30}$$
$$3$$

3.   Convert the mixed number $7\frac{3}{4}$ to an improper fraction.

   answer: $\frac{31}{4}$

First write 7 as $\frac{7}{1}$ and then convert this to an equivalent fraction with a denominator of 4 to match the denominator of $\frac{3}{4}$:

$$\frac{7}{1} = \frac{7 \cdot 4}{1 \cdot 4} = \frac{28}{4}$$

Now you know that 7 is the same as $\frac{28}{4}$, so $7\frac{3}{4} = \frac{28}{4} + \frac{3}{4}$. So if you have 28 quarters and add 3 more quarters, you'll have $28 + 3 = 31$ of these quarters. Thus, $7\frac{3}{4} = \frac{28}{4} + \frac{3}{4} = \frac{31}{4}$.

4.   Convert the mixed number $8\frac{1}{3}$ to an improper fraction.

   answer: $\frac{25}{3}$

First change 8 into a fraction with a denominator of 3:

$$8 = \frac{8}{1} = \frac{8 \cdot 3}{1 \cdot 3} = \frac{24}{3}$$

Now add:

$$8\frac{1}{3} = \frac{24}{3} + \frac{1}{3} = \frac{25}{3}$$

Work Problems

Use these problems to give yourself additional practice.

1. Convert the improper fraction $\frac{14}{6}$ to a mixed number.

2. Convert the improper fraction $\frac{100}{7}$ to a mixed number.

3. Convert the mixed number $1\frac{1}{4}$ to an improper fraction.

4. Convert the mixed number $10\frac{1}{2}$ to an improper fraction.

5. Convert the mixed number $3\frac{3}{8}$ to an improper fraction.

Worked Solutions

1. $2\frac{1}{3}$
    $$\begin{array}{r} 2 \\ 6\overline{)14} \\ -12 \\ \hline 2 \end{array}$$

    So $\frac{14}{6} = 2\frac{2}{6} = 2\frac{1}{3}$.

2. $14\frac{2}{7}$
    $$\begin{array}{r} 14 \\ 7\overline{)100} \\ -7 \\ \hline 30 \\ -28 \\ \hline 2 \end{array}$$

    So $\frac{100}{7} = 14\frac{2}{7}$.

3. $\frac{5}{4}$ $1\frac{1}{4} = \frac{4}{4} + \frac{1}{4} = \frac{4+1}{4} = \frac{5}{4}$

4. $\frac{21}{2}$ $10\frac{1}{2} = \frac{10\cdot2}{1\cdot2} + \frac{1}{2} = \frac{20+1}{2} = \frac{21}{2}$

5. $\frac{27}{8}$ $3\frac{3}{8} = \frac{3\cdot8}{1\cdot8} + \frac{3}{8} = \frac{24+3}{8} = \frac{27}{8}$

Adding and Subtracting Fractions

Some fractions are very easy to add or subtract, and, in fact, we've been doing it already. If you have two halves of a loaf of bread, you have a whole loaf of bread, so $\frac{1}{2} + \frac{1}{2} = \frac{2}{2} = 1$. This is all there is to it when the fractions both have the same denominator. The harder part is dealing with fractions that don't have the same denominator, in which case the best plan is to write them over a **common denominator**.

To add fractions:

1. Put both fractions over a common denominator.

2. Add the numerators over that common denominator.

To subtract fractions:

1. Put both fractions over a common denominator.

2. Subtract the numerators over that common denominator.

Example Problems

These problems show the answers and solutions.

1. Add $\frac{1}{5} + \frac{3}{5}$.

    answer: $\frac{4}{5}$

    Just add the numerators and write their sum over the common denominator:

    $$\frac{1}{5} + \frac{3}{5} = \frac{1+3}{5} = \frac{4}{5}$$

2. Add $\frac{3}{6} + \frac{2}{6}$.

    answer: $\frac{5}{6}$

    $$\frac{3}{6} + \frac{2}{6} = \frac{3+2}{6} = \frac{5}{6}$$

3. Add $\frac{1}{2} + \frac{1}{3}$.

    answer: $\frac{5}{6}$

    This time you don't start out with the same denominator, so it's a little more complicated. Notice, though, that $\frac{1}{2}$ is equivalent to $\frac{3}{6}$, and $\frac{1}{3}$ is equivalent to $\frac{2}{6}$, and you added those two fractions in the previous example. So, really you've already done this problem and can just write:

    $$\frac{1}{2} + \frac{1}{3} = \frac{3}{6} + \frac{2}{6} = \frac{3+2}{6} = \frac{5}{6}$$

4. Add $\frac{1}{4} + \frac{1}{5}$.

    answer: $\frac{9}{20}$

    Following the hint of the previous example, this would be easier if both fractions had the same denominator. Write both of them with a denominator of 20:

    $$\frac{1}{4} = \frac{1 \cdot 5}{4 \cdot 5} = \frac{5}{20}$$
    $$\frac{1}{5} = \frac{1 \cdot 4}{5 \cdot 4} = \frac{4}{20}$$

    So, now you can add them like this:

    $$\frac{1}{4} + \frac{1}{5} = \frac{5}{20} + \frac{4}{20} = \frac{5+4}{20} = \frac{9}{20}$$

5. $\frac{2}{3} + \frac{1}{8}$.

   answer: $\frac{19}{24}$

   You'll find a common denominator of 24 and add:

   $$\frac{2}{3} + \frac{1}{8} = \frac{2 \cdot 8}{3 \cdot 8} + \frac{1 \cdot 3}{8 \cdot 3} = \frac{16}{24} + \frac{3}{24} = \frac{19}{24}$$

6. Add $\frac{7}{10} + \frac{5}{6}$.

   answer: $\frac{23}{15}$

   Again, you'll find a common denominator and add. It would be possible to use 60 as your common denominator because both fractions easily could be written over 60. Your work would look like this:

   $$\frac{7}{10} + \frac{5}{6} = \frac{7 \cdot 6}{10 \cdot 6} + \frac{5 \cdot 10}{6 \cdot 10} = \frac{42}{60} + \frac{50}{60} = \frac{42 + 50}{60} = \frac{92}{60} = \frac{\cancel{2} \cdot \cancel{2} \cdot 23}{\cancel{2} \cdot \cancel{2} \cdot 3 \cdot 5} = \frac{23}{15}$$

   It would also be possible to choose other common denominators, like 30, in which case your work would look like this:

   $$\frac{7}{10} + \frac{5}{6} = \frac{7 \cdot 3}{10 \cdot 3} + \frac{5 \cdot 5}{6 \cdot 5} = \frac{21}{30} + \frac{25}{30} = \frac{21 + 25}{30} = \frac{46}{30} = \frac{\cancel{2} \cdot 23}{\cancel{2} \cdot 3 \cdot 5} = \frac{23}{15}$$

   Of course, the answer is the same either way, but the second way most people find a little easier because the numbers are smaller and a little easier to handle. 30 is called a **least common denominator** (often abbreviated **LCD**) for 10 and 6 because there is no smaller denominator into which both 10 and 6 divide evenly. Least common denominators can be found by comparing the prime factors of both denominators ($2 \cdot 5$ and $2 \cdot 3$ in this case) and including each of them at least as many times as they occur in either denominator ($2 \cdot 3 \cdot 5 = 30$ in this case).

7. $\frac{1}{2} - \frac{1}{3}$.

   answer: $\frac{1}{6}$

   Subtraction of fractions works just like addition, but once you have a common denominator, you subtract the numerators rather than add them:

   $$\frac{1}{2} - \frac{1}{3} = \frac{1 \cdot 3}{2 \cdot 3} - \frac{1 \cdot 2}{3 \cdot 2} = \frac{3}{6} - \frac{2}{6} = \frac{3 - 2}{6} = \frac{1}{6}$$

8. Subtract $\frac{13}{4} - \frac{4}{3}$.

   answer: $\frac{23}{12}$

   $$\frac{13}{4} - \frac{4}{3} = \frac{13 \cdot 3}{4 \cdot 3} - \frac{4 \cdot 4}{3 \cdot 4} = \frac{39 - 16}{12} = \frac{23}{12}$$

Work Problems

Use these problems to give yourself additional practice.

1. Add $\frac{2}{5} + \frac{1}{5}$.

2. Add $\frac{1}{3} + \frac{1}{4}$.

3. Add $\frac{6}{7} + \frac{3}{4}$.

4. Subtract $\frac{5}{3} - \frac{1}{3}$.

5. Subtract $\frac{11}{4} - \frac{5}{6}$.

Worked Solutions

1. $\frac{3}{5}$ $\quad \frac{2}{5} + \frac{1}{5} = \frac{1+2}{5} = \frac{3}{5}$

2. $\frac{7}{12}$ $\quad$ Use a common denominator of 12:

$$\frac{1}{3} + \frac{1}{4} = \frac{4 \cdot 1}{4 \cdot 3} + \frac{3 \cdot 1}{3 \cdot 4} = \frac{4}{12} + \frac{3}{12} = \frac{4+3}{12} = \frac{7}{12}$$

3. $\frac{45}{28}$ **or** $1\frac{17}{28}$ $\quad$ Use a common denominator of 28:

$$\frac{6}{7} + \frac{3}{4} = \frac{4 \cdot 6}{4 \cdot 7} + \frac{7 \cdot 3}{7 \cdot 4} = \frac{24+21}{28} = \frac{45}{28} \text{ or } 1\frac{17}{28}$$

4. $\frac{4}{3}$ **or** $1\frac{1}{3}$ $\quad \frac{5}{3} - \frac{1}{3} = \frac{5-1}{3} = \frac{4}{3}$ or $1\frac{1}{3}$

5. $\frac{23}{12}$ **or** $1\frac{11}{12}$ $\quad$ Use a common denominator of 12:

$$\frac{11}{4} - \frac{5}{6} = \frac{3 \cdot 11}{3 \cdot 4} - \frac{2 \cdot 5}{2 \cdot 6} = \frac{33-10}{12} = \frac{23}{12} \text{ or } 1\frac{11}{12}$$

Adding and Subtracting Mixed Numbers

It is always possible to add and subtract mixed numbers by first changing them to improper fractions and then adding or subtracting them using the procedures covered in the previous sections. Sometimes, however, it's easier to do it more directly, as the examples here will show.

Adding Mixed Numbers

First, we'll focus on addition.

Example Problems

These problems show the answers and solutions.

1. Add $1\frac{1}{2} + 3\frac{1}{4}$.

    answer: $4\frac{3}{4}$

    Write the numbers above one another like you did in Chapter 2 with the ones digits lined up and the fractions lined up, and then add down the columns. Also change the $\frac{1}{2}$ to an equivalent fraction with a denominator of 4:

    $$1\frac{2}{4}$$
    $$+3\frac{1}{4}$$
    $$\overline{4\frac{3}{4}}$$

2. Add $5\frac{7}{8} + 2\frac{1}{3}$.

    answer: $8\frac{5}{24}$

    Start by following the same procedure, using a common denominator of 24:

    $$5\frac{21}{24}$$
    $$+2\frac{8}{24}$$
    $$\overline{7\frac{29}{24}}$$

    Now, notice a drawback to this approach: It often leaves you with an improper fraction on a mixed number, which isn't very nice. So you have to do a little more work:

    $$7\frac{29}{24} = 7 + 1\frac{5}{24} = 8\frac{5}{24}$$

Subtracting Mixed Numbers

The trick when adding mixed numbers is that sometimes you have to do something similar to "carrying" as you did when adding whole numbers. So now when you get ready to subtract mixed numbers, expect sometimes to have to do something similar to "borrowing."

Example Problem

This problem shows the answer and solution.

1. Subtract $4\frac{1}{3} - 2\frac{5}{6}$.

    answer: $1\frac{1}{2}$

    Again line up the ones digits and the fraction parts. Use a common denominator of 6. The first step looks like the following:

$$4\frac{2}{6}$$
$$-2\frac{5}{6}$$

But when you go to subtract the fractions, $\frac{5}{6}$ is bigger than $\frac{2}{6}$, so you don't know how to subtract them. Borrow a 1 from the 4 and add this 1 to the $\frac{2}{6}$ to make it $\frac{8}{6}$. The rest looks like this:

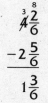

$$\overset{3}{4}\overset{8}{\frac{2}{6}}$$
$$-2\frac{5}{6}$$
$$1\frac{3}{6}$$

You still have to reduce the answer, so you end up with $4\frac{1}{3} - 2\frac{5}{6} = 1\frac{1}{2}$.

Work Problems

Use these problems to give yourself additional practice.

1. Add $4\frac{1}{4} + 3\frac{1}{4}$.

2. Add $1\frac{7}{8} + 8\frac{1}{5}$.

3. Add $3\frac{1}{3} + \frac{4}{5}$.

4. Subtract $5\frac{1}{2} - 2\frac{1}{4}$.

5. Subtract $9 - 3\frac{3}{4}$.

Worked Solutions

$$4\frac{1}{4}$$
1. $7\frac{1}{2}$ $+3\frac{1}{4}$
$$7\frac{1}{2}$$

2. $10\frac{3}{40}$ Use a common denominator of 40:

$$1\frac{35}{40}$$
$$+8\frac{8}{40}$$
$$9\frac{43}{40}$$

And then simplify this to $10\frac{3}{40}$.

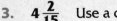

3. $4\frac{2}{15}$ Use a common denominator of 15:

$$3\frac{5}{15}$$
$$+\frac{12}{15}$$
$$3\frac{17}{15}$$

And then simplify this to $4\frac{2}{15}$.

4. $3\frac{1}{4}$ Use a common denominator of 4:

$$5\frac{2}{4}$$
$$-2\frac{1}{4}$$
$$3\frac{1}{4}$$

5. $5\frac{1}{4}$ Borrow 1 from the 9 and write that 1 as $\frac{4}{4}$:

$$\overset{8}{9}\frac{4}{4}$$
$$-3\frac{3}{4}$$
$$5\frac{1}{4}$$

Multiplying Fractions and Mixed Numbers

Multiplication and division of fractions and mixed numbers work very differently from addition and subtraction. You should pay close attention to the distinctions.

Multiplying Fractions

The procedure for multiplying fractions is simple.

To multiply fractions:

1. Multiply the numerators.
2. Multiply the denominators.
3. If necessary, reduce the answer.

Example Problems

These problems show the answers and solutions.

1. Multiply $\frac{4}{3} \cdot \frac{7}{2}$.

   answer: $\frac{14}{3}$

Multiply the 4 and 7 from the numerators and multiply the 3 and 2 from the denominators, like this:

$$\frac{4}{3} \cdot \frac{7}{2} = \frac{4 \cdot 7}{3 \cdot 2} = \frac{28}{6}$$

And then notice that this can be reduced to $\frac{14}{3}$. Some people would also simplify this to $4\frac{2}{3}$, but it's best to use your judgment on this because in a situation where you're going to do more after this, it might be easier to keep an improper fraction than a mixed number.

2. Multiply $5 \times \frac{7}{10}$.

   answer: $3\frac{1}{2}$

   You need to write the 5 as a fraction first and then multiply and simplify:

   $$\frac{5}{1} \cdot \frac{7}{10} = \frac{35}{10} = 3\frac{5}{10} = 3\frac{1}{2}$$

   It's also possible to make this a little bit easier by canceling common factors of the numerator and denominator before you multiply, like this:

   $$\frac{5}{1} \cdot \frac{7}{10} = \frac{\cancel{5} \cdot 7}{2 \cdot \cancel{5}} = \frac{7}{2} = 3\frac{1}{2}$$

3. Multiply $\frac{3}{8} \cdot \frac{10}{9}$.

   answer: $\frac{5}{12}$

   To make things simpler, we can cancel common factors and then multiply:

   $$\frac{3}{8} \cdot \frac{10}{9} = \frac{\cancel{3} \cdot \cancel{2} \cdot 5}{\cancel{2} \cdot 2 \cdot 2 \cdot \cancel{3} \cdot 3} = \frac{5}{12}$$

Multiplying Mixed Numbers

The best way to multiply mixed numbers is to change them to improper fractions and then multiply those exactly as you have been.

Example Problems

These problems show the answers and solutions.

1. Multiply $3\frac{1}{2} \times 5\frac{1}{4}$.

   answer: $\frac{147}{8}$ or $18\frac{3}{8}$

Convert $3\frac{1}{2}$ to $\frac{7}{2}$ and $5\frac{1}{4}$ to $\frac{21}{4}$, and then multiply:

$$\frac{7}{2} \cdot \frac{21}{4} = \frac{147}{8} \text{ or } 18\frac{3}{8}$$

2. Multiply $4\frac{1}{3} \times 9$.

answer: 39

Convert the mixed number to an improper fraction, write the 9 as $\frac{9}{1}$, and then multiply.

$$\frac{13}{\overset{}{\underset{1}{\cancel{3}}}} \cdot \frac{\overset{3}{\cancel{9}}}{1} = \frac{39}{1} = 39$$

Notice that the cancellation is written a little bit differently here—we didn't actually write out the factorization, we just made notes about what terms were left after we cancelled the common factors.

Work Problems

Use these problems to give yourself additional practice.

1. Multiply $\frac{3}{2} \cdot \frac{5}{3}$.

2. Multiply $\frac{7}{12} \cdot \frac{8}{5}$.

3. Multiply $3\frac{1}{4} \times \frac{1}{2}$.

4. Multiply $7 \cdot \frac{1}{7}$.

5. Multiply $3\frac{1}{3} \times 3\frac{1}{3}$.

Worked Solutions

1. $\dfrac{5}{2}$ $\dfrac{\overset{1}{\cancel{3}}}{2} \cdot \dfrac{5}{\underset{1}{\cancel{3}}} = \dfrac{5}{2}$

2. $\dfrac{14}{15}$ $\dfrac{7}{\underset{3}{\cancel{12}}} \cdot \dfrac{\overset{2}{\cancel{8}}}{5} = \dfrac{14}{15}$

3. $\dfrac{13}{8}$ $\dfrac{13}{4} \cdot \dfrac{1}{2} = \dfrac{13}{8}$

4. 1 $\dfrac{\overset{1}{\cancel{7}}}{1} \cdot \dfrac{1}{\underset{1}{\cancel{7}}} = \dfrac{1}{1} = 1$

5. $\dfrac{100}{9}$ $\dfrac{10}{3} \cdot \dfrac{10}{3} = \dfrac{100}{9}$

Dividing Fractions and Mixed Numbers

The procedures for dividing fractions and mixed numbers are very similar to those for multiplying.

Dividing Fractions

To divide fractions, take the first fraction and multiply it by the reciprocal of the second fraction.

Remember that the reciprocal is obtained by interchanging the numerator and denominator of a fraction.

Example Problems

These problems show the answers and solutions.

1. Divide $\frac{1}{2} \div \frac{1}{4}$.

    answer: 2

    First find the reciprocal of the second fraction, which would be $\frac{4}{1}$, and then multiply:

    $$\frac{1}{2} \div \frac{1}{4} = \frac{1}{2} \times \frac{4}{1} = \frac{1}{\cancel{2}_1} \times \frac{\cancel{4}^2}{1} = \frac{2}{1} = 2$$

2. Divide $\frac{7}{8} \div \frac{3}{4}$.

    answer: $\frac{7}{6}$

    The reciprocal of $\frac{3}{4}$ is $\frac{4}{3}$, so:

    $$\frac{7}{8} \div \frac{3}{4} = \frac{7}{\cancel{8}_2} \times \frac{\cancel{4}^1}{3} = \frac{7}{6}$$

    And, as in the previous section, you don't necessarily want to change this to a mixed number, so leave it like this.

Dividing Mixed Numbers

As with multiplication, division of mixed numbers is easiest if you first change them to improper fractions. Then divide by multiplying by the reciprocal, just as before.

Example Problems

These problems show the answers and solutions.

1. $1\frac{1}{4} \div 3\frac{1}{2}$.

    answer: $\frac{5}{14}$

    Convert the fractions to $\frac{5}{4}$ and $\frac{7}{2}$ and then multiply $\frac{5}{4}$ by the reciprocal of $\frac{7}{2}$:

    $$\frac{5}{4} \div \frac{7}{2} = \frac{5}{\overset{}{\underset{2}{4}}} \times \frac{\overset{1}{2}}{7} = \frac{5}{14}$$

2. Divide $2\frac{1}{7} \div 5\frac{1}{6}$.

    answer: $\frac{90}{217}$

    Convert to improper fractions and multiply by the reciprocal:

    $$\frac{15}{7} \div \frac{31}{6} = \frac{15}{7} \times \frac{6}{31} = \frac{90}{217}$$

Work Problems

Use these problems to give yourself additional practice.

1. Divide $\frac{1}{3} \div \frac{5}{6}$.

2. Divide $\frac{4}{7} \div \frac{6}{5}$.

3. Divide $1\frac{1}{5} \div \frac{4}{3}$.

4. Divide $4\frac{1}{3} \div 3\frac{1}{12}$.

5. Divide $1 \div \frac{3}{4}$.

Worked Solutions

1. $\frac{2}{5}$ $\frac{1}{3} \div \frac{5}{6} = \frac{1}{\overset{}{\underset{}{3}}} \times \frac{\overset{2}{6}}{5} = \frac{2}{5}$

2. $\frac{10}{21}$ $\frac{4}{7} \div \frac{6}{5} = \frac{\overset{2}{4}}{7} \times \frac{6}{5} = \frac{10}{21}$

3. $\frac{9}{10}$ $\frac{6}{5} \div \frac{4}{3} = \frac{\overset{3}{6}}{5} \times \frac{3}{\overset{}{\underset{2}{4}}} = \frac{9}{10}$

4. $\frac{52}{37}$ $\frac{13}{3} \div \frac{37}{12} = \frac{13}{\overset{}{\underset{}{3}}} \times \frac{\overset{4}{12}}{37} = \frac{52}{37}$

5. $\frac{4}{3}$ $\frac{1}{1} \div \frac{3}{4} = \frac{1}{1} \times \frac{4}{3} = \frac{4}{3}$

Compound Fractions

When either the numerator or the denominator of a fraction contains another fraction, it is called a **compound fraction** (or sometimes a **complex fraction**, but this is confusing because it has nothing to do with the complex numbers, which you can learn about later in algebra). You can simplify compound fractions by first simplifying the numerator and denominator separately, writing each as an improper fraction, and then performing division on the results.

Example Problems

These problems show the answers and solutions.

1. Simplify $\dfrac{1\frac{1}{4}}{5\frac{1}{3}}$.

    answer: $\dfrac{15}{64}$

    In this case there is no simplification to be done in either the numerator or the denominator, so just write them as mixed numbers $\frac{5}{4}$ and $\frac{16}{3}$ and then multiply $\frac{5}{4}$ by the reciprocal of $\frac{16}{3}$:

    $$\frac{1\frac{1}{4}}{5\frac{1}{3}} = \frac{\frac{5}{4}}{\frac{16}{3}} = \frac{5}{4} \div \frac{16}{3} = \frac{5}{4} \times \frac{3}{16} = \frac{15}{64}$$

2. Simplify $\dfrac{2 - \frac{1}{2}}{2 + \frac{1}{2}}$.

    answer: $\dfrac{3}{5}$

    First work within the numerator and the denominator separately, writing them as improper fractions and then performing the division:

    $$\frac{2 - \frac{1}{2}}{2 + \frac{1}{2}} = \frac{\frac{4}{2} - \frac{1}{2}}{\frac{4}{2} + \frac{1}{2}} = \frac{\frac{3}{2}}{\frac{5}{2}} = \frac{3}{2} \times \frac{2}{5} = \frac{3}{5}$$

3. Simplify $\dfrac{1}{1 + \dfrac{1}{1 + \frac{1}{3}}}$.

    answer: $\dfrac{4}{7}$

    Start in the denominator, first by simplifying the fraction that appears in it:

    $$\frac{1}{1 + \dfrac{1}{1 + \frac{1}{3}}} = \frac{1}{1 + \dfrac{1}{\frac{4}{3}}} = \frac{1}{1 + \frac{3}{4}} = \frac{1}{\frac{4}{4} + \frac{3}{4}} = \frac{1}{\frac{7}{4}} = 1 \times \frac{4}{7} = \frac{4}{7}$$

Work Problems

1. Simplify $\dfrac{1 + \dfrac{1}{3}}{2}$.

2. Simplify $\dfrac{\dfrac{1}{2} - \dfrac{1}{3}}{\dfrac{1}{4}}$.

3. Simplify $\dfrac{1}{1 + \dfrac{1}{1 + \dfrac{1}{2}}}$.

Worked Solutions

1. $\dfrac{2}{3}$ $\dfrac{1 + \dfrac{1}{3}}{2} = \dfrac{\dfrac{3}{3} + \dfrac{1}{3}}{2} = \dfrac{\dfrac{4}{3}}{2} = \dfrac{4}{3} \times \dfrac{1}{2} = \dfrac{2}{3}$

2. $\dfrac{2}{3}$ $\dfrac{\dfrac{1}{2} - \dfrac{1}{3}}{\dfrac{1}{4}} = \dfrac{\dfrac{3}{6} - \dfrac{2}{6}}{\dfrac{1}{4}} = \dfrac{\dfrac{1}{6}}{\dfrac{1}{4}} = \dfrac{1}{6} \times \dfrac{4}{1} = \dfrac{2}{3}$

3. $\dfrac{3}{5}$ $\dfrac{1}{1 + \dfrac{1}{1 + \dfrac{1}{2}}} = \dfrac{1}{1 + \dfrac{1}{\dfrac{3}{2}}} = \dfrac{1}{1 + \dfrac{2}{3}} = \dfrac{1}{\dfrac{3}{3} + \dfrac{2}{3}} = \dfrac{1}{\dfrac{5}{3}} = 1 \times \dfrac{3}{5} = \dfrac{3}{5}$

Chapter 4
Decimals

Decimals (sometimes called decimal numbers) are another way (different from fractions) to represent parts of things, like a temperature of ninety-eight and six-tenths degrees, or a price of three dollars and ninety-five cents.

Place Value

Decimal numbers extend the place value system, using digits to the right of the decimal point to represent tenths, then hundredths, thousandths, and so forth in the same way digits to the left of the decimal place represent ones, then tens, hundreds, thousands, and so forth. Since one tenth $= 10^{-1}$, one hundredth $= 10^{-2}$, and so on, this fits nicely with the system for whole numbers we talked about at the beginning of Chapter 2. So 856.723 means 8 hundreds, 5 tens, 6 ones, 7 tenths, 2 hundredths, and 3 thousandths. See how this fits in the grid of Table 4-1, and compare it to Table 2-1 where we did the same thing for the whole number 856.

Table 4-1						
Thousands	**Hundreds**	**Tens**	**Ones**	**Tenths**	**Hundredths**	**Thousandths**
1000	100	10	1	0.1	0.01	0.001
10^3	10^2	10^1	10^0	10^{-1}	10^{-2}	10^{-3}
	8	5	6	7	2	3

Notice that it's customary to write a zero in the ones place, like "0.03," when writing three hundredths. Otherwise, the decimal point may be easily overlooked or mistaken for multiplication written with a dot.

As with whole numbers, we can break a decimal number up to show the place value of each digit.

Example Problems

These problems show the answers and solutions.

1. Write 354.7 in expanded notation.

    answer: $(3 \times 10^2) + (5 \times 10^1) + (4 \times 10^0) + (7 \times 10^{-1})$

    $$354.7 = (3 \times 100) + (5 \times 10) + (4 \times 1) + (7 \times 0.1)$$
    $$= (3 \times 10^2) + (5 \times 10^1) + (4 \times 10^0) + (7 \times 10^{-1})$$

2. Write 0.0057 in expanded notation.

    answer: $(0 \times 10^0) + (0 \times 10^{-1}) + (0 \times 10^{-2}) + (5 \times 10^{-3}) + (7 \times 10^{-4})$

    $0.0057 = (0 \times 1) + (0 \times 0.1) + (0 \times 0.01) + (5 \times 0.001) + (7 \times 0.0001)$

    $= (0 \times 10^0) + (0 \times 10^{-1}) + (0 \times 10^{-2}) + (5 \times 10^{-3}) + (7 \times 10^{-4})$

Work Problems

Use these problems to give yourself additional practice.

1. Write 84.79 in expanded notation.

2. Write 600.08 in expanded notation.

3. Write 2.718 in expanded notation.

Worked Solutions

1. $(8 \times 10^1) + (4 \times 10^0) + (7 \times 10^{-1}) + (9 \times 10^{-2})$

    $84.79 = (8 \times 10) + (4 \times 1) + (7 \times 0.1) + (9 \times 0.01)$

    $= (8 \times 10^1) + (4 \times 10^0) + (7 \times 10^{-1}) + (9 \times 10^{-2})$

2. $(6 \times 10^2) + (0 \times 10^1) + (0 \times 10^0) + (0 \times 10^{-1}) + (8 \times 10^{-2})$

    $600.08 = (6 \times 100) + (0 \times 10) + (0 \times 1) + (0 \times 0.1) + (8 \times 0.01)$

    $= (6 \times 10^2) + (0 \times 10^1) + (0 \times 10^0) + (0 \times 10^{-1}) + (8 \times 10^{-2})$

3. $(2 \times 10^0) + (7 \times 10^{-1}) + (1 \times 10^{-2}) + (8 \times 10^{-3})$

    $2.718 = (2 \times 1) + (7 \times 0.1) + (1 \times 0.01) + (8 \times 0.001)$

    $= (2 \times 10^0) + (7 \times 10^{-1}) + (1 \times 10^{-2}) + (8 \times 10^{-3})$

Comparing Decimals

Comparing two decimals is easy if you write them both with the same number of decimal places.

Example Problems

These problems show the answers and solutions.

1. Which is greater, 0.5 or 0.37?

answer: 0.5   Since 0.37 has two decimal places, you need to write 0.5 with two decimal places also. You write a zero in the hundredths place, 0.50. Now you're comparing two numbers that both represent parts out of one hundred; so since fifty hundredths is more than thirty-seven hundredths, you know that 0.5 is larger than 0.37.

2.   Arrange the numbers 1.34, 1.6, and 1.097 in order from smallest to largest.

answer: 1.097, 1.34, and 1.6   First you write the numbers as 1.340, 1.600, and 1.097. Now thinking of these as 1097 thousandths, 1340 thousandths, and 1600 thousandths, you see that they should be arranged in order from smallest to largest as 1.097, 1.34, and 1.6.

Work Problems

Use these problems to give yourself additional practice.

1.   Which is greater, 0.9 or 0.66?

2.   Which is greater, 3.14 or 2.78?

3.   Arrange the numbers 0.034, 0.12, and 0.4 in order from smallest to largest.

4.   Arrange the numbers 1.001, 0.02, and 0.3 in order from smallest to largest.

Worked Solutions

1.   **0.9**   You write the numbers with two decimal places: 0.90 and 0.66. Since 90 hundredths is more than 66 hundredths, you know that 0.9 is greater than 0.66.

2.   **3.14**   The numbers are already written with two decimal places; so since 314 hundredths is more than 278 hundredths, you know that 3.14 is greater than 2.78.

3.   **0.034, 0.12, and 0.4**   Since the most decimal places in any of the three numbers is three, you write all three numbers with three decimal places, as 0.034, 0.120, and 0.400. Then, since 34 thousandths is less than 120 thousandths, which in turn is less than 400 thousandths, you see that the numbers are already in order from smallest to largest.

4.   **0.02, 0.3, and 1.001**   Since the most decimal places in any of these numbers is three, you write them each with three decimal places as 1.001, 0.020, and 0.300. Then since 20 thousandths is less than 300 thousandths, which in turn is less than 1001 thousandths, you arrange the numbers in order from smallest to largest as 0.02, 0.3, and 1.001.

Decimal Computation

You can add, subtract, multiply, and divide decimal numbers just as you did whole numbers in Chapter 2. In most cases, the procedures for these operations are a lot like the ones for whole numbers, and the only extra work involves keeping track of where the decimal point is in the answers.

Adding Decimals

To add two or more decimal numbers, write them one above the other just as you did with whole numbers (Chapter 2), and add starting with the rightmost column. When you write the numbers, you need to line up the ones, tens, and so forth just like you did before. With decimals, this means lining up the decimal points one above the other.

Example Problems

These problems show the answers and solutions.

1. Add 45.6 + 103.2.

    answer: 148.8

    Notice that you put the decimal point in the answer directly below the decimal point in the two numbers you add:

    $$\begin{array}{r} 45.6 \\ +103.2 \\ \hline 148.8 \end{array}$$

2. Add 2.75 + 1.5.

    answer: 4.25

    Write the 1.5 as 1.50 to keep the columns straight and carry where necessary:

    $$\begin{array}{r} \overset{1}{2}.75 \\ +1.50 \\ \hline 4.25 \end{array}$$

3. Add 675.18 + 8.3 + 5.99.

    answer: 689.47

    Remember to line up the decimal points and write 8.3 as 8.30:

    $$\begin{array}{r} \overset{1\ 1\ 1}{675}.18 \\ 8.30 \\ +5.99 \\ \hline 689.47 \end{array}$$

4. Add 93.1538 + 0.06.

    answer: 93.2138

    Write the 0.06 as 0.0600 to keep the columns lined up:

    $$\begin{array}{r} \overset{1}{9}3.1538 \\ +0.0600 \\ \hline 93.2138 \end{array}$$

Work Problems

Use these problems to give yourself additional practice.

1. Add 6.3 + 1.4.

2. Add 23.4 + 19.6.

3. Add 99.3 + 23.8.

4. Add 7.03 + 4.55 + 2.38.

5. Add 303.03 + 707.07.

Worked Solutions

1.
$$\begin{array}{r} 6.3 \\ +1.4 \\ \hline 7.7 \end{array}$$

2.
$$\begin{array}{r} \overset{1\,1}{23.4} \\ +19.6 \\ \hline 43.0 \end{array}$$

3.
$$\begin{array}{r} \overset{1\,1}{99.3} \\ +23.8 \\ \hline 123.1 \end{array}$$

4.
$$\begin{array}{r} \overset{1}{7.03} \\ 4.55 \\ +2.38 \\ \hline 13.96 \end{array}$$

5.
$$\begin{array}{r} \overset{1}{3}0\overset{1}{3}.03 \\ +707.07 \\ \hline 1010.10 \end{array}$$

Subtracting Decimals

Subtracting decimals is also just like subtracting natural numbers (Chapter 2), again with the requirement that you keep the decimal points lined up and put a decimal point in your answer in a corresponding position.

Example Problems

These problems show the answers and solutions.

1. Subtract 4.3 − 1.2.

   answer: 3.1

Line up the decimal points:

$$\begin{array}{r} 4.3 \\ -\ 1.2 \\ \hline 3.1 \end{array}$$

2. Subtract 9.8 − 4.9.

answer: 4.9

Line up the decimal points and borrow just like with natural numbers:

$$\begin{array}{r} \overset{8}{\cancel{9}}.\overset{1}{8} \\ -\ 4.9 \\ \hline 4.9 \end{array}$$

3. Subtract 302.5 − 13.8.

answer: 288.7

Again, you line up the decimal places and borrow like with natural numbers:

$$\begin{array}{r} \overset{2}{\cancel{3}}\overset{9}{\cancel{0}}\overset{11}{\cancel{2}}.\overset{1}{5} \\ -\ 1\,3\,.\,8 \\ \hline 2\,88.\,7 \end{array}$$

Work Problems

Use these problems to give yourself additional practice.

1. Subtract 10.3 − 3.2.

2. Subtract 8.3 − 2.3.

3. Subtract 17.08 − 0.95.

4. Subtract 100 − 26.7.

5. Subtract 55.555 − 7.777.

Worked Solutions

1. **7.1**
$$\begin{array}{r} 10.3 \\ -\ 3.2 \\ \hline 7.1 \end{array}$$

2. **6.0**
$$\begin{array}{r} 8.3 \\ -\ 2.3 \\ \hline 6.0 \end{array}$$

3.   **16.13** $\overset{\;\;\overset{6}{1}7.\overset{1}{.}08}{\underline{-0.95}}$
                  16.13

4.   **73.3** $\overset{\;\;\overset{9}{\cancel{1}}\overset{9}{\cancel{0}}\cancel{0}.\overset{1}{0}}{\underline{-26.7}}$
                  73.3

5.   **47.778** $\overset{\;\;\overset{4}{\cancel{5}}\overset{14}{\cancel{5}}.\overset{14}{\cancel{5}}\overset{14}{\cancel{5}}5}{\underline{-7.777}}$
                  4.778

Multiplying Decimals

Multiplying decimals follows exactly the same plan as multiplying whole numbers (Chapter 2), except for the decimal point itself. The number of decimal places in the answer should be equal to the sum of the number of decimal places in the two numbers being multiplied (and you add leading zeros if necessary). Thus, if a number with one digit to the right of the decimal point is multiplied by another number with two digits to the right of the decimal place, the answer should have three digits to the right of the decimal place. In practice it can be slightly more complicated than this (see Example Problem 1 that follows), but the best way to understand it is by following along with some examples.

Example Problems

These problems show the answers and solutions.

1.   Multiply 3.4×2.7.

     answer: 9.18

     The first stage will look like this:

$$
\begin{array}{r}
3.4 \\
\times 2.7 \\
\hline
238 \\
+68 \\
\hline
918
\end{array}
$$

Next you count the number of digits to the right of the decimal points in the two numbers you're multiplying. Since there's one digit to the right of the decimal point in each, for two digits total, you insert the decimal point in your answer two places from the right:

$$
\begin{array}{r}
3.4 \\
\times 2.7 \\
\hline
238 \\
+68 \\
\hline
9.18
\end{array}
$$

So $3.4 \times 2.7 = 9.18$. Notice that this answer is reasonable: You multiplied a number a little more than 3 and a number a little less than 3, and got an answer close to 9. It's always a good idea to make sure that your answers are reasonable this way because putting the decimal point in the wrong place can lead to outrageous answers.

2. Multiply 7.5×3.

    answer: 22.5

    First you multiply by the usual procedure, and then you put the decimal point one digit from the right in your answer, since there was a total of one digit to the right of the decimal points in the two numbers you were multiplying:

$$\begin{array}{r} 7.5 \\ \times\,3 \\ \hline 22.5 \end{array}$$

So $7.5 \times 3 = 22.5$, which is reasonable since it's close to $7 \times 3 = 21$.

3. Multiply 2.4×21.8.

    answer: 52.32

    Multiply and put the decimal point two digits from the right in your answer. Also notice that you switch the two numbers, since it's easier to carry out the multiplication if the second number has fewer digits. (The answer is the same because of the commutative property of multiplication, from Chapter 1.)

$$\begin{array}{r} 21.8 \\ \times\;\;2.4 \\ \hline 872 \\ +436 \\ \hline 52.32 \end{array}$$

So $21.8 \times 2.4 = 52.32$. Notice that this is reasonable, since $2 \times 22 = 44$, and this is a little bigger than that.

4. Multiply 0.02×0.038.

    answer: 0.00076

    Multiply and put the decimal point five digits from the right in the answer—notice that this requires adding zeros to the left in your answer:

$$\begin{array}{r} 0.038 \\ \times 0.02 \\ \hline 0.00076 \end{array}$$

5. Multiply 352.5×5.2.

    answer: 1833.00

Multiply and put the decimal point two digits from the right in your answer:

$$
\begin{array}{r}
352.5 \\
\times 5.2 \\
\hline
70\ 50 \\
1762\ 5 \\
\hline
1833.00
\end{array}
$$

So $352.5 \times 5.2 = 1833.00$. Notice that you could also write your answer 1833, leaving out the zeros to the right of the decimal point, but it's very important to keep those zeros until you've properly located the decimal point. The answer 18.33 would not be reasonable, since $5 \times 300 = 1500$ and your answer should certainly be larger than this.

Work Problems

Use these problems to give yourself additional practice.

1. Multiply 3.3×4.1.

2. Multiply 13.6×7.

3. Multiply 0.06×1.2.

4. Multiply 0.00318×0.05.

5. Multiply 4.67×17.2.

Worked Solutions

1. **13.53**
$$
\begin{array}{r}
3.3 \\
\times 4.1 \\
\hline
33 \\
+132 \\
\hline
13.53
\end{array}
$$

2. **95.2**
$$
\begin{array}{r}
13.6 \\
\times 7 \\
\hline
95.2
\end{array}
$$

3. **0.072**   Either:
$$
\begin{array}{r}
0.06 \\
\times 1.2 \\
\hline
12 \\
+6 \\
\hline
0.072
\end{array}
\qquad \text{or} \qquad
\begin{array}{r}
1.2 \\
\times 0.06 \\
\hline
0.072
\end{array}
$$

4. **0.0001590**
$$\begin{array}{r} 0.00318 \\ \times\,0.05 \\ \hline 0.0001590 \end{array}$$

5. **80.324**
$$\begin{array}{r} 4.67 \\ \times\,17.2 \\ \hline 934 \\ 3269 \\ +467 \\ \hline 80.324 \end{array}$$

Dividing Decimals

Dividing decimals is essentially like dividing natural numbers (Chapter 2), but with careful handling of the decimal points, which is a little bit more complicated than what you've seen with the other operations. The simplest possibility is if the divisor happens to be a natural number, in which case you just need to line up the columns in a way similar to what you've done before.

Example Problems

These problems show the answers and solutions.

1. Divide $17.5 \div 5$.

    answer: 3.5   You work just as you did with whole numbers and put the decimal point in your answer directly above the decimal point in your dividend:

$$\begin{array}{r} 3.5 \\ 5\overline{)\,17.5} \\ -15 \\ \hline 25 \\ -25 \\ \hline 0 \end{array}$$

    Remember this also means that $3.5 \times 5 = 17.5$, so you can tell that your answer is reasonable.

    Now if the divisor is a decimal, make it a whole number by moving the decimal point over all the way to the right and moving the decimal point in the dividend just as many places to the right. This works because of the way in which you multiply decimals—see the comment at the end of the next example:

2. Divide $17.52 \div 2.4$.

    answer: 7.3   Since the divisor, 2.4, has one decimal place, move the decimal point in both numbers one digit to the right. This means that you're really working out $175.2 \div 24$, like this:

$$
\begin{array}{r}
7.3 \\
24\overline{)175.2} \\
-168 \\
\hline
72 \\
-72 \\
\hline
0
\end{array}
$$

So $17.52 \div 2.4 = 7.3$ because $175.2 \div 24 = 7.3$, which means $7.3 \times 24 = 175.2$, or $7.3 \times 2.4 = 17.52$. Notice that both of these multiplication problems are essentially the same, just with different numbers of decimal places.

So far the divisions have come out evenly, with zeros after the final subtraction. This doesn't always happen.

3. Divide $3.65 \div 0.4$.

   answer: 9.125   First we move the decimal points in both numbers over one place to the right, so we work out $36.5 \div 4$:

$$
\begin{array}{r}
9.1 \\
4\overline{)36.5} \\
-36 \\
\hline
05 \\
-4 \\
\hline
1
\end{array}
$$

Notice that the result of this last subtraction is not a zero. What you do next is write another zero to the right in your dividend and continue:

$$
\begin{array}{r}
9.12 \\
4\overline{)36.50} \\
-36 \\
\hline
05 \\
-4 \\
\hline
10 \\
-8 \\
\hline
2
\end{array}
$$

Again, you haven't ended up with a zero at the bottom yet. You can write another zero in the dividend and see what happens:

$$
\begin{array}{r}
9.125 \\
4\overline{)36.500} \\
-36 \\
\hline
05 \\
-4 \\
\hline
10 \\
-8 \\
\hline
20 \\
-20 \\
\hline
0
\end{array}
$$

This time you got a zero, and so $3.65 \div 0.4 = 9.125$.

4. Divide $1.55 \div 0.3$.

   answer: 5.16666666...   First move the decimal points in both numbers over one place to the right, so that you work out $15.5 \div 3$:

$$
\begin{array}{r}
5.1 \\
3\overline{)15.5} \\
-15 \\
\hline
05 \\
-3 \\
\hline
2
\end{array}
$$

See that the result of your last subtraction is not a zero. What you do next is write another zero in the dividend and continue:

$$
\begin{array}{r}
5.16 \\
3\overline{)15.50} \\
-15 \\
\hline
05 \\
-3 \\
\hline
20 \\
-18 \\
\hline
2
\end{array}
$$

Because you still didn't end up with zero, you can repeat the process of putting a zero to the right of the current dividend and working out the next digit in your quotient:

$$
\begin{array}{r}
5.166 \\
3\overline{)15.500} \\
-15 \\
\hline
05 \\
-3 \\
\hline
20 \\
-18 \\
\hline
2
\end{array}
$$

You could repeat the process again, but you really should notice the pattern that has developed instead. Every time you work out another digit in your answer it will be a 6, and the remainder passed to the next stage will again be a 2. So $1.55 \times 0.3 = 5.16666666...$, with the 6s repeating forever. These repeating decimals (sometimes called nonterminating decimals) are sometimes written as shown previously, with the "..." indicating that the pattern continues, or written as $5.1\overline{6}$, with the bar over the 6, meaning that digit repeats forever.

Also notice that you can put a bar over more than one digit if necessary, so $0.\overline{24}$ means $0.24242424...$ with the "24" repeating forever, and $0.\overline{142857}$ means $0.142857142857...$ with the block "142857" repeating forever.

Work Problems

1. Divide $1.25 \div 5$.

2. Divide $6 \div 0.15$.

3. Divide $35.6 \div 0.25$.

4. Divide $30.4 \div 1.2$.

5. Divide $1.354 \div 0.09$.

Worked Solutions

1. **0.25**
$$
\begin{array}{r}
.25 \\
5\overline{)\,1.25} \\
-1\,0 \\
\hline
25 \\
-25 \\
\hline
0
\end{array}
$$

2. **40** Move the decimal point two places to the right (inserting zeros where necessary) and divide $600 \div 15$:

$$
\begin{array}{r}
40 \\
15\overline{)\,600} \\
-60 \\
\hline
00 \\
-0 \\
\hline
0
\end{array}
$$

3. **142.4**   Move the decimal point two places to the right (inserting zeros where necessary) and divide $3560 \div 25$:

$$
\begin{array}{r}
142.4 \\
25 \overline{)\ 3560.0} \\
-25 \\
\hline
106 \\
-100 \\
\hline
60 \\
-50 \\
\hline
100 \\
-100 \\
\hline
0
\end{array}
$$

4. **25.$\overline{3}$**   Move the decimal point one place to the right and work out $304 \div 12$:

$$
\begin{array}{r}
25.33 \\
12 \overline{)\ 304.00} \\
-24 \\
\hline
64 \\
-60 \\
\hline
40 \\
-36 \\
\hline
40 \\
-36 \\
\hline
4
\end{array}
$$

So $30.4 \div 1.2 = 25.\overline{3}$.

5. **15.0$\overline{4}$**   We move the decimal points two places to the right and work out $135.4 \div 9$:

$$
\begin{array}{r}
15.04 \\
9 \overline{)\ 135.40} \\
-9 \\
\hline
45 \\
-45 \\
\hline
04 \\
-0 \\
\hline
40 \\
-36 \\
\hline
4
\end{array}
$$

So $1.354 \div 0.09 = 15.0\overline{4}$.

Estimating

Working out exact answers to problems with decimals can be extremely time-consuming. If only an approximate answer is needed, you can round off the numbers you're working with (usually to whole numbers) and get approximate answers much more easily.

To **round off** a decimal,

1. Underline the value of the digit to which you are rounding off.

2. Look one place to the right of the underlined digit.

3. If this digit is a 5 or higher, increase the underlined value by 1 and change all the numbers to the right of this value to zeros. If the digit is less than 5, leave the underlined digit as it is and change all the digits to its right to zero.

4. If there are zeros to the right of the decimal place at the end of the number, drop them.

Example Problems

These problems show the answers and solutions.

1. Round 56.38 to the nearest whole number.

   answer: 56

   We underline the ones digit, in this case the 6. Then, because the digit to the right of the 6 is less than 5, 5$\underline{6}$.38 is rounded down to 56.00, or just 56.

2. Round 135.685 to the nearest tenth.

   answer: 135.7

   Underline the tenths digit, in this case the 6. Then 135.$\underline{6}$85 is rounded up to 135.700 or just 135.7.

3. Estimate 2.36 + 5.9 by first rounding to the nearest whole numbers.

   answer: 8

   $2.36 + 5.9 \approx 2 + 6 = 8$

4. Estimate 93.4 − 5.9 by rounding to the nearest whole numbers.

   answer: 87

   $93.4 - 5.9 \approx 93 - 6 = 87$

5. Estimate 23.1 × 4.8 by rounding to the nearest whole numbers.

   answer: 115

   $23.1 \times 4.8 \approx 23 \times 5 = 115$

6. Estimate 123.4567 ÷ 2.942 by rounding to the nearest whole numbers.

    answer: 41

    $123.4567 \div 2.942 \approx 123 \div 3 = 41$

Work Problems

1. Round 67.48 to the nearest whole number.

2. Estimate 7.56 + 2.403 by rounding to the nearest whole numbers.

3. Estimate 7.56 − 2.403 by rounding to the nearest whole numbers.

4. Estimate 7.56 × 2.403 by rounding to the nearest whole numbers.

5. Estimate 7.56 ÷ 2.403 by rounding to the nearest whole numbers.

Worked Solutions

1. **67**   Underline the ones digit, and 6$\underline{7}$.48 rounds down to 67.

2. **10** $7.56 + 2.403 \approx 8 + 2 = 10$

3. **6** $7.56 - 2.403 \approx 8 - 2 = 6$

4. **16** $7.56 \times 2.403 \approx 8 \times 2 = 16$

5. **4** $7.56 \div 2.403 \approx 8 \div 2 = 4$

Changing between Fractions and Decimals

Fractions and decimals both represent numbers between whole numbers. Sometimes you might prefer to work with decimals, and other times with fractions, so you need to know how to change back and forth.

Changing Fractions to Decimals

To change a fraction to a decimal, just work out the division. Many fractions become repeating decimals.

Example Problems

These problems show the answers and solutions.

1. Change $\frac{1}{8}$ to a decimal.

    answer: 0.125

$\frac{1}{8}$ means $1 \div 8$, so we work out:

$$
\begin{array}{r}
0.125 \\
8\overline{)\,1.000} \\
-0 \\
\hline
10 \\
-8 \\
\hline
20 \\
-16 \\
\hline
40 \\
-40 \\
\hline
0
\end{array}
$$

2. Change $\frac{1}{3}$ to a decimal.

   answer: $0.\overline{3}$

   $\frac{1}{3}$ means $1 \div 3$, so we work out:

$$
\begin{array}{r}
0.33 \\
3\overline{)\,1.00} \\
-0 \\
\hline
10 \\
-9 \\
\hline
10 \\
-9 \\
\hline
1
\end{array}
$$

3. Change $\frac{7}{9}$ to a decimal.

   answer: $0.\overline{7}$.

   $\frac{7}{9}$ means $7 \div 9$, so we work out:

$$
\begin{array}{r}
0.77 \\
9\overline{)\,7.00} \\
-0 \\
\hline
70 \\
-63 \\
\hline
70 \\
-63 \\
\hline
7
\end{array}
$$

Work Problems

Use these problems to give yourself additional practice.

1. Change $\frac{3}{5}$ to a decimal.

2. Change $\frac{1}{25}$ to a decimal.

3. Change $\frac{1}{6}$ to a decimal.

4. Change $\frac{4}{9}$ to a decimal.

5. Change $\frac{16}{99}$ to a decimal.

Worked Solutions

1. **0.6** $\frac{3}{5}$ means $3 \div 5$, so we work out:

$$
\begin{array}{r}
0.6 \\
5 \overline{)\ 3.0} \\
\underline{-0} \\
30 \\
\underline{-30} \\
0
\end{array}
$$

2. **0.04** $\frac{1}{25}$ means $1 \div 25$, so we work out:

$$
\begin{array}{r}
0.04 \\
25 \overline{)\ 1.00} \\
\underline{-0} \\
10 \\
\underline{-0} \\
100 \\
\underline{-100} \\
0
\end{array}
$$

3. **.1$\overline{6}$** $\frac{1}{6}$ means $1 \div 6$, so we work out:

$$
\begin{array}{r}
0.16 \\
6 \overline{)\ 1.00} \\
\underline{-0} \\
10 \\
\underline{-6} \\
40 \\
\underline{-36} \\
4
\end{array}
$$

4. **0.$\overline{4}$** $\frac{4}{9}$ means $4 \div 9$, so we work out:

$$
\begin{array}{r}
0.44 \\
9 \overline{\smash{)}\ 4.00} \\
\underline{-0} \\
40 \\
\underline{-36} \\
40 \\
\underline{-36} \\
4
\end{array}
$$

5. **.$\overline{16}$** $\frac{16}{99}$ means $16 \div 99$, so we work out:

$$
\begin{array}{r}
0.1616 \\
99 \overline{\smash{)}\ 16.0000} \\
\underline{-0} \\
160 \\
\underline{-99} \\
610 \\
\underline{-594} \\
160 \\
\underline{-99} \\
610 \\
\underline{-594} \\
16
\end{array}
$$

Changing Decimals to Fractions

There are two possibilities when you change a decimal to a fraction. The first is that the decimal terminates (that is, it doesn't repeat). In this case, the place-value system really does the work for us.

Example Problems

1. Change 0.4 into a fraction.

    answer: $\frac{4}{10}$ 0.4 has a 4 in the tenths place, so 4 tenths can be written as $\frac{4}{10}$. This can also be reduced to $\frac{2}{5}$.

2. Change 0.681 to a fraction.

    answer: $\frac{681}{1000}$ 0.681 is 681 thousandths, or $\frac{681}{1000}$, which can't be reduced further.

3. Change 3.14 to a fraction.

   answer: $3\frac{14}{100}$ 3.14 is 3 and 14 hundredths, or $3\frac{14}{100} = 3\frac{7}{50}$.

   The other possibility is trying to change a decimal that repeats to a fraction. The key in this case is an observation you might have made while doing the last set of work problems—that $\frac{1}{9} = 0.\bar{1}$, $\frac{2}{9} = 0.\bar{2}$, and so forth.

4. Change $0.\bar{5}$ to a fraction.

   answer: $\frac{5}{9}$ We can just recognize that $0.\bar{5}$ is the decimal form of $\frac{5}{9}$.

   If two or more digits repeat, we can change to a fraction by writing those digits over 99, or 999, or as many nines as there are digits repeating.

5. Change $0.\overline{15}$ to a fraction.

   answer: $\frac{5}{33}$ Since two digits are repeating, we write them over 99, so $0.\overline{15} = \frac{15}{99} = \frac{5}{33}$.

6. Change $0.58\bar{3}$ to a fraction.

   answer: $\frac{7}{12}$ This time there's only one digit repeating, but it didn't start in the tenths place, so our job is a little bit more complicated. We'll handle this in parts. First, 0.58 is 58 hundredths, or $\frac{58}{100}$ (we could reduce this, but in these problems it's often better not to until the last stage). Second, the rest of the decimal, $0.00\bar{3}$, is the same as $0.\bar{3}$ (which we know is $\frac{3}{9} = \frac{1}{3}$) but divided by 100, so $0.00\bar{3} = \frac{1}{3} \div 100 = \frac{1}{300}$. Finally we put these two parts together:

   $$0.58\bar{3} = \frac{58}{100} + \frac{1}{300} = \frac{174}{300} + \frac{1}{300} = \frac{175}{300} = \frac{7}{12}$$

   This covers both the possibility of decimals that terminate and decimals that don't terminate, but repeat. There is another possibility, which is decimals that don't terminate and don't repeat either. These decimals include things like $\sqrt{2} \approx 1.41421356237$ and $\pi \approx 3.14159265359$, and can't be changed to fractions. They are called irrational numbers, precisely because they can't be expressed as the ratio of two integers.

Work Problems

Use these problems to give yourself additional practice.

1. Change 0.28 to a fraction.

2. Change 5.5 to a fraction.

3. Change $3.\bar{2}$ to a fraction.

4. Change $0.\overline{06}$ to a fraction.

5. Change $0.208\bar{3}$ to a fraction.

Worked Solutions

1. $\frac{7}{25}$ This is 28 hundredths, or $\frac{28}{100} = \frac{7}{25}$.

2. $5\frac{1}{2}$ This is 5 and 5 tenths, or $5\frac{5}{10} = 5\frac{1}{2}$.

3. $3\frac{2}{9}$ Since the digit 2 is repeating, we write that part as $0.\overline{2} = \frac{2}{9}$, so then $3.\overline{2} = 3\frac{2}{9}$.

4. $\frac{2}{33}$ Since there are two digits repeating, we write them over 99 as $\frac{06}{99} = \frac{2}{33}$.

5. $\frac{5}{24}$ We can write 0.208 as $\frac{208}{1000}$, and the other $0.000\overline{3}$ is $\frac{3}{9} = \frac{1}{3}$ but with the decimal point moved three places over, so divided by 1000, $0.000\overline{3} = \frac{1}{3} \div 1000 = \frac{1}{3000}$. Then putting these together, we have $0.208\overline{3} = \frac{208}{1000} + \frac{1}{3000} = \frac{624}{3000} + \frac{1}{3000} = \frac{625}{3000} = \frac{5}{24}$.

Chapter 5
Percents

Percent means "out of one hundred," so a percent can be treated as a fraction with a denominator of 100. For instance, $67\% = \frac{67}{100}$, and $\frac{1}{10} = \frac{10}{100} = 10\%$.

Changing Percents, Decimals, and Fractions

Many quantities can be expressed either as decimals, fractions, or percents. For instance, the fraction $\frac{1}{2}$ is the same as the decimal 0.5 or 50%. You can change between decimals, fractions, and percents easily with step-by-step procedures.

Changing Decimals to Percents

To change decimals to percents, do the following:

1. Move the decimal point two places to the right.
2. Put a percent sign after the number.

Example Problems

These problems show the answers and solutions.

1. Change 0.47 to a percent.

   answer: 47%

   First move the decimal two places to the right and then add a percent sign:

   $0.47 \rightarrow 47. \rightarrow 47\%$

2. Change 1.75 to a percent:

   answer: 175%

   First move the decimal two places to the right and then add a percent sign:

   $1.75 \rightarrow 175. \rightarrow 175\%$

Work Problems

Use these problems to give yourself additional practice.

1.   Change 0.32 to a percent.

2.   Change 0.07 to a percent.

3.   Change 6.29 to a percent.

4.   Change 0.008 to a percent.

Worked Solutions

1.   **32%**   By following the steps, you have 0.32 → 32. → 32%.

2.   **7%**   By following the steps, you have 0.07 → 07. → 7%.

3.   **629%**   By following the steps, you have 6.29 → 629. → 629%.

4.   **0.8%**   By following the steps, you have 0.008 → 00.8 → 0.8%.

Changing Percents to Decimals

To change percents to decimals, do the following:

1. Take the percent sign off.
2. Move the percent sign two places to the left.
3. If necessary, add zeros.

Example Problems

These problems show the answers and solutions.

1.   Change 53% to a decimal.

   answer: 0.53

   First eliminate the percent sign.

   53% → 53

   Then move the decimal point two places to the left.

   53 → 0.53

2.   Change 6% to a decimal.

   answer: 0.06

   Using the steps provided, you get:

$6\% \rightarrow 6 \rightarrow 0.0\underset{\smile}{6}$

Note that it was necessary here to add a zero when we moved the decimal point.

Work Problems
Use these problems to give yourself additional practice.

1. Change 44% to a decimal.

2. Change 17% to a decimal.

3. Change 187% to a decimal.

4. Change 3% to a decimal.

5. Change 0.9% to a decimal.

Worked Solutions

1. **0.44** $44\% \rightarrow 44 \rightarrow 0.\underset{\smile}{44}$

2. **0.17** $17\% \rightarrow 17 \rightarrow 0.\underset{\smile}{17}$

3. **1.87** $187\% \rightarrow 187 \rightarrow 1.\underset{\smile}{87}$

4. **0.03** $3\% \rightarrow 3 \rightarrow 0.0\underset{\smile}{3}$

5. **0.009** $0.9\% \rightarrow 0.9 \rightarrow 0.00\underset{\smile}{9}$

Changing Fractions to Percents
To change fractions to percents, do the following:

1. Change the fraction to a decimal.
2. Change the decimal to a percent.

Example Problems
These problems show the answers and solutions.

1. Change $\frac{1}{4}$ to a percent.

    answer: 25%

    First change the fraction to a decimal: $\frac{1}{4} = 0.25$.

    Then follow the steps to change the decimal to a percent: $0.25 \rightarrow 2\underset{\smile}{5}. \rightarrow 25\%$.

2. Change $\frac{7}{2}$ to a percent.

   answer: 350%

   First, change it to a decimal and then to a percent:

   $\frac{7}{2} = 3.5 \rightarrow 350. \rightarrow 350\%$

Work Problems
Use these problems to give yourself additional practice.

1. Change $\frac{1}{2}$ to a percent.

2. Change $\frac{3}{5}$ to a percent.

3. Change $\frac{7}{20}$ to a percent.

4. Change $\frac{3}{4}$ to a percent.

5. Change $\frac{1}{25}$ to a percent.

Worked Solutions

1. $\frac{1}{2} = 0.5 \rightarrow 50. \rightarrow 50\%$

2. $\frac{3}{5} = 0.6 \rightarrow 60. \rightarrow 60\%$

3. $\frac{7}{20} = 0.35 \rightarrow 35. \rightarrow 35\%$

4. $\frac{3}{4} = 0.75 \rightarrow 75. \rightarrow 75\%$

5. $\frac{1}{25} = 0.04 \rightarrow 04. \rightarrow 4\%$

Changing Percents to Fractions
To change percents to fractions, do the following:

1. Drop the percent sign.
2. Write it as a fraction over 100.
3. Reduce if necessary.

Example Problems
These problems show the answers and solutions.

1. Change 21% to a fraction.

   answer: $\frac{21}{100}$

   First drop the percent sign: 21% → 21.

Then write as a fraction over 100: $21 \to \frac{21}{100}$.

Because it cannot be reduced, you are finished.

2. Change 125% to a fraction.

answer: $1\frac{1}{4}$

First drop the percent sign: $125\% \to 125$.

Then write as a fraction over 100: $125 \to \frac{125}{100}$.

Now reduce: $\frac{125}{100} = 1\frac{25}{100} = 1\frac{1}{4}$.

Work Problems

Use these problems to give yourself additional practice.

1. Change 32% to a fraction.

2. Change 9% to a fraction.

3. Change 75% to a fraction.

4. Change 139% to a fraction.

5. Change 245% to a fraction.

Worked Solutions

1. $\frac{8}{25}$ $32\% \to 32 \to \frac{32}{100} \to \frac{8}{25}$

2. $\frac{9}{100}$ $9\% \to 9 \to \frac{9}{100}$

3. $\frac{3}{4}$ $75\% \to 75 \to \frac{75}{100} \to \frac{3}{4}$

4. $1\frac{39}{100}$ $139\% \to 139 \to \frac{139}{100} \to 1\frac{39}{100}$

5. $2\frac{9}{10}$ $245\% \to 245 \to \frac{245}{100} \to \frac{49}{20} \to 2\frac{9}{20}$

Applications of Percents

Percents are helpful in solving many practical problems. Here is a selection of some of their most common uses.

Finding Percents of a Number

To find a certain percentage of some given number, change the percent to a decimal or fraction (whichever you prefer) and multiply. In these problems, it can help to think of the word "of" as meaning multiply.

Example Problems

These problems show the answers and solutions.

1. What is 40% of 80?

   answer: 32

   (Using decimals) Change the percent to a decimal:

   $40\% \to 0.40$

   Then multiply:

   $0.40 \cdot 80 = 32$

   (Using fractions) Change the percent to a fraction:

   $40\% \to \frac{2}{5}$

   Then multiply:

   $\frac{2}{5} \cdot 80 = \frac{160}{5} = 32$

2. What is 15% of 60?

   answer: 9

   (Using decimals) Change the percent to a decimal:

   $15\% \to 0.15$

   Then multiply:

   $0.15 \cdot 60 = 9$

   (Using fractions) Change the percent to a fraction:

   $15\% \to \frac{15}{100} \to \frac{3}{20}$

   Then multiply:

   $\frac{3}{20} \cdot 60 = \frac{180}{20} = 9$

3. What is $\frac{3}{4}\%$ of 24?

   answer: 0.18 or $\frac{9}{50}$

   (Using decimals) Change the percent to a decimal:

   $\frac{3}{4}\% \to 0.75\% \to 0.0075$

   Then multiply:

   $0.0075 \cdot 24 = 0.18$

(Using fractions) Change the percent to a fraction:

$$\frac{3}{4}\% \rightarrow \frac{3}{4} \cdot \frac{1}{100} \rightarrow \frac{3}{400}$$

Then multiply:

$$\frac{3}{400} \cdot 24 = \frac{72}{400} = \frac{9}{50}$$

Work Problems

Use these problems to give yourself additional practice.

1.   What is 30% of 90?

2.   What is 50% of 128?

3.   What is 20% of 70?

4.   What is $\frac{1}{2}$% of 15?

5.   What is 150% of 30?

Worked Solutions

1.   **27**   (Using decimals) 30% = 0.30 · 90 = 27

   (Using fractions) 30% = $\frac{30}{100} = \frac{3}{10} \cdot 90 = \frac{270}{10} = 27$

2.   **64**   (Using decimals) 50% = 0.50 · 128 = 64

   (Using fractions) 50% = $\frac{50}{100} = \frac{1}{2} \cdot 128 = 64$

3.   **14**   (Using decimals) 20% = 0.20 · 70 = 14

   (Using fractions) 20% = $\frac{20}{100} = \frac{1}{5} \cdot 70 = 14$

4.   **0.075 or $\frac{3}{40}$**   (Using decimals) $\frac{1}{2}$% = 0.5% = 0.005 · 15 = 0.075

   (Using fractions) $\frac{1}{2}$% = $\frac{1}{200} \cdot 15 = \frac{3}{40}$

5.   **45**   (Using decimals) 150% = 1.50 · 30 = 45

   (Using fractions) 150% = $\frac{150}{100} = \frac{3}{2} \cdot 30 = \frac{90}{2} = 45$

Finding What Percent One Number Is of Another Number

One way to find what percent one number is of another number is the division method. To use this method, take the number after the word *of* and divide it into the number next to the word *is*. Then change the answer to a percent.

An alternative method to find what percent one number is of another is to use the equation method. For this method, turn the question into an equation one word at a time. (For help on solving simple equations, see Chapter 12.) For *what*, substitute the variable *x*; for *is*, substitute an *equal sign (=)*; for the word *of*, substitute a *multiplication sign (·)*. Change percents to decimals or fractions, whichever is easier for you. Then solve the equation.

Example Problems

These problems show the answers and solutions.

Use the division method to solve problems 1 and 2:

1. 24 is what percent of 80?

   answer: $30\frac{1}{3}\%$

   $\frac{24}{80} = \frac{3}{10} = 0.3 = 30\%$

2. 20 is what percent of 60?

   answer: $33.\overline{3}\%$ or $33\frac{1}{3}\%$

   $\frac{20}{60} = \frac{1}{3} = 0.33\overline{3} = 33.\overline{3}\%$ or $33\frac{1}{3}\%$

Use the division method to solve problems 3 and 4:

3. 10 is what percent of 40?

   answer: 25%

   $$10 = x(40)$$
   $$\frac{10}{40} = \frac{x(40)}{40}$$
   $$\frac{10}{40} = x$$
   $$\frac{1}{4} = x$$
   $$0.25 = x$$
   $$25\% = x$$

4. 15 is what percent of 75?

   answer: 20%

   $$15 = x(75)$$
   $$\frac{15}{75} = \frac{x(75)}{75}$$
   $$\frac{15}{75} = x$$
   $$\frac{1}{5} = x$$
   $$0.2 = x$$
   $$20\% = x$$

Work Problems

Use these problems to give yourself additional practice.

Use the division method to solve problems 1 and 2:

1. 12 is what percent of 48?

2. 20 is what percent of 50?

Use the equation method to solve problems 3 and 4:

3. 35 is what percent of 105?

4. 14 is what percent of 7?

Worked Solutions

1. **25%** $\frac{12}{48} = \frac{1}{4} = 0.25 = 25\%$

2. **40%** $\frac{20}{50} = \frac{2}{5} = 0.4 = 40\%$

3. $33\frac{1}{3}\%$

$$35 = x(105)$$

$$\frac{35}{105} = \frac{x105}{105}$$

$$\frac{35}{105} = x$$

$$\frac{1}{3} = x$$

$$33\frac{1}{33}\% = x$$

4. **200%**

$$14 = x(7)$$

$$\frac{14}{7} = \frac{x(7)}{7}$$

$$\frac{14}{7} = x$$

$$2 = x$$

$$200\% = x$$

Finding a Number When a Percent of It Is Known

The division method can also be used to find a number when a percent of it is known. To apply this method, take the percent, change it into a decimal or a fraction (whichever you prefer), and divide that number into the other number.

An alternative approach to finding a number when a percent of it is known is to use the equation method. For this method, turn the question into an equation one word at a time. (For help on solving simple equations, see Chapter 12). For *what*, substitute the variable x; for *is*, substitute an *equal sign (=)*; for the word *of*, substitute a *multiplication sign (·)*. Change percents to decimals or fractions, whichever is easier for you. Then solve the equation.

Example Problems

These problems show the answers and solutions.

1. 30 is 50% of what number?

   answer: 60

   (Using decimals) 50% = 0.50, and $\frac{30}{0.50} = 60$

   (Using fractions) 50% = $\frac{1}{2}$, and $\frac{30}{1} \div \frac{1}{2} = \frac{30 \cdot 2}{1} = 60$

2. 15 is 75% of what number?

   answer: 20

   (Using decimals) 75% = 0.75, and $\frac{15}{0.75} = 20$

   (Using fractions) 75% = $\frac{3}{4}$, and $\frac{15}{1} \div \frac{3}{4} = \frac{15}{1} \cdot \frac{4}{3} = 5 \cdot 4 = 20$

3. 40 is 25% of what number?

   answer: 160

$$\text{(Using decimals) } 40 = (0.25)x$$
$$\frac{40}{0.25} = \frac{(0.25)x}{0.25}$$
$$\frac{40}{0.25} = x$$
$$160 = x$$
$$\text{(Using fractions) } 40 = \left(\frac{1}{4}\right)x$$
$$4 \cdot 40 = \left(\frac{1}{4}\right)x \cdot 4$$
$$160 = x$$

4. 60 is 80% of what number?

   answer: 75

$$\text{(Using decimals) } 60 = (0.80)x$$
$$\frac{60}{0.80} = \frac{(0.80)x}{0.80}$$
$$\frac{60}{0.80} = x$$
$$75 = x$$

$$(\text{Using fractions}) \ 60 = \frac{4}{5}x$$

$$\frac{5}{4} \cdot 60 = \frac{5}{4} \cdot \frac{4}{5} \cdot x$$

$$5 \cdot 15 = x$$

$$75 = x$$

Work Problems

Use these problems to give yourself additional practice.

Use the division method to solve problems 1–3.

1. 24 is 40% of what number?

2. 20 is 50% of what number?

3. 16 is 60% of what number?

Use the equation method to solve problems 4 and 5.

4. 15 is 25% of what number?

5. 45 is 30% of what number?

Worked Solutions

1. 60

(Using decimals) $\frac{24}{0.4} = 60$

(Using fractions) $\frac{24}{1} \div \frac{2}{5} = 24 \cdot \frac{5}{2} = 60$

2. 40

(Using decimals) $\frac{20}{0.5} = 40$

(Using fractions) $\frac{20}{1} \div \frac{1}{2} = \frac{20 \cdot 2}{1} = 40$

3. $26\frac{2}{3}$

(Using decimals) $\frac{16}{0.6} = 26\frac{2}{3}$

(Using fractions) $\frac{16}{1} \div \frac{3}{5} = \frac{16 \cdot 5}{3} = \frac{80}{3} = 26\frac{2}{3}$

4. **60**

(Using decimals) $15 = (0.25)x$

$$\frac{15}{0.25} = \frac{(0.25)x}{0.25}$$

$$\frac{15}{0.25} = x$$

$$60 = x$$

(Using fractions) $15 = \left(\frac{1}{4}\right)x$

$$15 \cdot 4 = x$$

$$60 = x$$

5. **150**

(Using decimals) $45 = (0.30)x$

$$\frac{45}{0.30} = \frac{(0.30)x}{0.30}$$

$$\frac{45}{0.30} = x$$

$$150 = x$$

(Using fractions) $45 = 45 \cdot \frac{10}{3} = x$

$$150 = x$$

The Proportion Method

Another easy way commonly used to solve any of the three basic types of percent problems is the proportion method. First set up an empty proportion $\left(\frac{?}{?} = \frac{?}{?}\right)$ and then insert the appropriate values by following these steps:

1. Take the percent, or x if the percent is unknown and write it as a fraction over 100 on the left side of the proportion.

2. Take the base number (usually the number following the word *of*) and put it in the denominator on the right side of the proportion.

3. Take the other number, the one representing some portion of the base (often the number next to the word *is*) and put it in the numerator on the right side of the proportion.

4. Now solve for the unknown x.

Example Problems

These problems show the answers and solutions.

1. 20 is what percent of 50?

   answer: 40%

Set up a blank proportion:

$$\frac{?}{?} = \frac{?}{?}$$

Step 1: $\frac{x}{100} = \frac{?}{?}$

Step 2: $\frac{x}{100} = \frac{?}{50}$

Step 3: $\frac{x}{100} = \frac{20}{50}$

$$\frac{x}{100} = \frac{2}{5}$$

Step 4: $x = \frac{2}{5} \cdot \frac{100}{1}$

$$x = 40\%$$

2. 60 is 20% of what number?

answer: 300

This time you have 20%, or 20 out of 100, corresponding to 60 out of some unknown quantity that we'll label x, so write:

$$\frac{20}{100} = \frac{60}{x}$$

$$\frac{1}{5} = \frac{60}{x}$$

$$x = 60 \cdot 5$$

$$x = 300$$

3. What number is 30% of 90?

answer: 27

You want to know what number, x, out of 90 corresponds to 30%, or 30 out of 100, so:

$$\frac{30}{100} = \frac{x}{90}$$

$$\frac{3}{10} = \frac{x}{90}$$

$$3 \cdot 90 = x \cdot 10$$

$$27 = x$$

Work Problems

Use these problems to give yourself additional practice.

1. 40 is what percent of 150?

2. What number is 90% of 200?

3. 24 is 60% of what number?

4. What number is $66\frac{2}{3}\%$ of 30?

Worked Solutions

1. **$26\frac{2}{3}$**   You want to know what x out of 100 corresponds to 40 out of 150, so:

$$\frac{x}{100} = \frac{40}{150}$$
$$\frac{x}{100} = \frac{4}{15}$$
$$x \cdot 15 = 4 \cdot 100$$
$$x = \frac{80}{3}$$
$$x = 26\frac{2}{3}$$

2. **180**   You want to know what x out of 200 corresponds to 90 out of 100, so:

$$\frac{90}{100} = \frac{x}{200}$$
$$\frac{9}{10} = \frac{x}{200}$$
$$1800 = 10x$$
$$180 = x$$

3. **40**   You want to know an x for which 24 out of x corresponds to 60 out of 100, so:

$$\frac{60}{100} = \frac{24}{x}$$
$$\frac{3}{5} = \frac{24}{x}$$
$$3x = 120$$
$$x = 40$$

4. **20**   You want to know an x for which x out of 30 corresponds to $66\frac{2}{3}$ out of 100, so:

$$\frac{66\frac{2}{3}}{100} = \frac{x}{30}$$
$$30 \cdot 66\frac{2}{3} = 100x$$
$$2000 = 100x$$
$$20 = x$$

Finding Percent Increase or Percent Decrease

Often, we want to know by what percent something increased or decreased. There's a simple formula for finding percent change, whether it's increase or decrease.

To find percent change, use the following formula:

$$\text{percent change} = \frac{\text{change}}{\text{starting value}}$$

Example Problems

These problems show the answers and solutions.

1. What is the percent decrease of a $600 item on sale for $500?

    answer: $16\frac{2}{3}\%$

    The change in the price is $600 – $500 = $100, and the starting value was $600, so the percent change $= \frac{100}{600} = 16\frac{2}{3}\%$.

2. If the population of Mexico was 67.4 million in 1980 and 76.6 million in 1985, what was the percent increase in population over this period?

    answer: 13.6%

    The change in population is 76.6 – 67.4 = 9.2 million, and the starting population was 67.4 million, so the percent change $= \frac{9.2}{67.4} \approx 13.6\%$.

Work Problems

Use these problems to give yourself additional practice.

1. What is the percent change from 80 to 120?

2. What is the percent decrease of a $900 item on sale for $700?

3. What is the percent increase in pay per month if someone making $1420 each month gets a raise to $1462.60 each month?

4. What is the percent change if gasoline prices go from $1.299 a gallon to $1.699 a gallon?

Worked Solutions

1. **50%**   The change is 120 – 80 = 40, and the starting value was 80, so the percent change $= \frac{40}{80} = 50\%$.

2. **$22\frac{2}{9}\%$**   The change is $900 – $700 = 200, and the starting value was $900, so the percent change $= \frac{\$200}{\$900} = \frac{2}{9} = 22\frac{2}{9}\%$.

3. **3%**   The change is $1462.60 – $1420 = $42.60, and the starting value was $1420, so the percent change $= \frac{\$42.6}{\$1420} = 0.03 = 3\%$.

4. **30.8%**   The change is $1.699 – $1.299 = $0.400 a gallon, and the starting value was $1.299 a gallon, so the percentage change $\frac{\$0.40}{\$1.299} \approx .308 = 30.8\%$.

Chapter 6
Integers and Rationals

Integers

Think of the group of numbers called **integers** as the collection of the positive and negative natural numbers, along with zero. Examples of integers would be 7, −5, 1000, −777, and 0. Another way to obtain all the integers is to take the whole numbers as well as the negatives of the whole numbers.

Number Lines

You can place the integers in order on a number line. Use 0, which is sometimes called the **origin** because it's the starting point, as the middle of the number line. Then, put the positive integers on the right of zero and the negative integers on the left of zero:

$$\xleftarrow{\hspace{2cm}} \quad -4 \quad -3 \quad -2 \quad -1 \quad 0 \quad 1 \quad 2 \quad 3 \quad 4 \quad \xrightarrow{\hspace{2cm}}$$

When looking at two different numbers on a number line, the larger number will always be to the right, no matter what the number's sign is.

Adding Integers

Think of adding two integers as starting on the number line at the mark corresponding to the first integer and then moving from there by the amount of the second integer—going to the right if the second integer is positive and to the left if the second integer is negative. As you read the following examples, it might help to think about the corresponding positions on the number line.

The usual procedures for adding integers are simple and can be stated with two rules:

1. To add two integers that are either both positive or both negative, add the two numbers (ignoring their signs) and then place the sign that the original numbers had on that answer.

2. To add two integers that do not have the same sign (one is positive and one is negative), take the number that's bigger (ignoring the signs) and subtract the number that's smaller (still ignoring the signs). Then maintain the sign of the number farthest from zero on the number line.

Example Problems

These problems show the answers and solutions.

1. Add

$$+2$$
$$+\ +4$$

answer: +6

Both of these integers are positive, so add 2 and 4 and put a positive sign on the result.

2. Add

$$-7$$
$$+\ -4$$

answer: −11

Both of these integers are negative, so add 7 and 4 and put a negative sign on the result.

3. Add

$$+3$$
$$+\ -11$$

answer: −8

Because these two numbers have different signs, take $11 - 3 = 8$, and put a negative sign on the 8 (negative because −11 is farther from 0 on the number line than +3 is).

4. Add

$$-6$$
$$+\ +9$$

answer: +3

Because these two numbers have different signs, take $9 - 6 = 3$, and put a positive sign on the 3 (positive because +9 is farther from 0 on the number line than is −6).

When adding integers, you can also set them up horizontally.

5. Add +5 + +12.

answer: 17

$$+5 + +12 = +17$$

6. Add +8 + −9.

answer: −1

$$+8 + -9 = -1$$

Work Problems

Use these problems to give yourself additional practice.

1. Add

$$+13$$
$$+\ +8$$

2. Add

$$-10$$
$$+\ +4$$

3. Add $+7 + -6$.

4. Add $-3 + -11$.

Worked Solutions

1. **+21**   The answer is positive because both numbers are positive.

2. **−6**   The answer is negative because -10 is farther from zero on the number line than is $+4$.

3. **+1**   The answer is positive because $+7$ is farther from zero on the number line than is -6.

4. **−14**   The answer is negative because both numbers are negative.

Subtracting Integers

You also can think of subtracting integers along a number line. Start out at the first number and then move backward by the second number—to the right for a negative number or to the left for a positive number. Again, you might want to think about each of the following examples on a number line as well as work them out symbolically.

The rule for subtracting integers is

> To subtract integers, change the sign of the second number and then add the two numbers using the rules for adding integers.

Example Problems

These problems show the answers and solutions.

1. Subtract $\begin{array}{r} +12 \\ -+7 \\ \hline \end{array}$.

   answer: $+5$

   Change the problem to adding the opposite of the second number.

$$\begin{array}{r} +12 \\ +-7 \\ \hline +5 \end{array}$$

2. Subtract $\begin{array}{r} -14 \\ --11 \\ \hline \end{array}$.

   answer: -3

   Change the problem to adding the opposite of the second number.

$$\begin{array}{r} -14 \\ ++11 \\ \hline -3 \end{array}$$

3. Subtract $\begin{array}{r} +10 \\ -\ -4 \\ \hline \end{array}$.

    answer: $+14$

    Change the problem to adding the opposite of the second number.

    $$\begin{array}{r} +10 \\ +\ +4 \\ \hline +14 \end{array}$$

4. Subtract $\begin{array}{r} -16 \\ -\ +6 \\ \hline \end{array}$.

    answer: -22

    Change the problem to adding the opposite of the second number.

    $$\begin{array}{r} -16 \\ +\ -6 \\ \hline -22 \end{array}$$

As in addition, you can also set subtraction problems up horizontally.

5. Subtract $+4 - -6$.

    answer: 10

    $$+4 - -6 = +4 + +6 = +10$$

6. Subtract $-17 - -4$.

    answer: -13

    $$-17 - -4 = -17 + +4 = -13$$

Work Problems

Use these problems to give yourself additional practice.

1. Subtract $\begin{array}{r} +9 \\ -\ +4 \\ \hline \end{array}$.

2. Subtract $\begin{array}{r} -13 \\ -\ +11 \\ \hline \end{array}$.

3. Subtract $\begin{array}{r} -10 \\ -\ -2 \\ \hline \end{array}$.

4. Subtract $+8 - -3$.

5. Subtract $-9 - -6$.

Worked Solutions

1. **+5** $\quad\begin{array}{r} +9 \\ +-4 \\ \hline +5 \end{array}$

2. **−24** $\quad\begin{array}{r} -13 \\ +-11 \\ \hline -24 \end{array}$

3. **−8** $\quad\begin{array}{r} -10 \\ ++2 \\ \hline -8 \end{array}$

4. **+11** $+8 - -3 = +8 + +3 = +11$

5. **−3** $-9 - -6 = -9 + +6 = -3$

Minus Preceding Grouping Symbols

When you have a minus sign preceding a parenthesis or other grouping symbol, it means that you are subtracting what is inside the parentheses from what comes before it. To do this, change the sign of each integer inside the parentheses and add.

Note: If no sign is in front of a number, the number is positive.

Example Problems

These problems show the answers and solutions.

1. Subtract $7 - (+4 - 6 + 8 - 3)$.

    answer: 4

    Change the − to a + and reverse the signs inside the parentheses.

    $$7 - (+4 - 6 + 8 - 3) = 7 + (-4 + 6 - 8 + 3) = 7 + (-3) = 4$$

2. Subtract $18 - (+5 - 7 + 67)$.

    answer: −47

    Change the − to a + and reverse the signs inside the parentheses.

    $$18 - (+5 - 7 + 67) = 18 + (-5 + 7 - 67) = 18 + (-65) = -47$$

    Another way of solving a problem with a minus preceding a parenthesis is to add all of the positive numbers within the parentheses. Next, combine that with the sum of all the negative numbers inside the parentheses. Finally, subtract.

3. Subtract $18 - (+5 - 7 + 67)$.

   answer: -47

   $$18 - (+5 - 7 + 67) = 18 - (+72 - 7) = 18 - (+65) = -47$$

4. Subtract $6 - (3 - 4 + 2 - 9)$.

   answer: 14

   $$6 - (3 - 4 + 2 - 9) = 6 - (5 - 13) = 6 - (-8) = 14$$

Work Problems

Use these problems to give yourself additional practice.

1. Subtract $5 - (+2 - 3 + 7)$.

2. Subtract $20 - (-6 + 4 + 9)$.

3. Subtract $4 - (12 - 4 + 5)$.

4. Subtract $16 - (-3 + 9 - 6 + 1)$.

5. Subtract $11 - (7 - 13 + 4)$.

Worked Solutions

1. **−1** $5 - (+2 - 3 + 7) = 5 + (-2 + 3 - 7) = 5 + (-6) = -1$

2. **13** $20 - (-6 + 4 + 9) = 20 - (-6 + 13) = 20 - (7) = 13$

3. **−9** $4 - (12 - 4 + 5) = 4 + (-12 + 4 - 5) = 4 + (-13) = -9$

4. **15** $16 - (-3 + 9 - 6 + 1) = 16 - (10 - 9) = 16 - (1) = 15$

5. **13** $11 - (7 - 13 + 4) = 11 + (-7 + 13 - 4) = 11 + (2) = 13$

Multiplying and Dividing Integers

To multiply or divide integers, follow the same steps as you would when dealing with normal numbers. When you are finished, look back at how many negative signs were in the original problem. If the original problem had an even number of negative signs, the answer is positive; if the original problem had an odd number of negative signs, the answer is negative. These rules might seem strange, but actually they're just a summary of what you've already done: $2 \times (-3)$ means start at -3 on a number line and then go 3 steps left from there because $2 \times (-3)$ is the same as $-3 + -3$. So, a positive times a negative means repeating a move to the left on a number line and leaves you left of zero for a negative answer. Similar reasoning applies to all the other combinations of signs.

You need to remember the following:

❑ The product or quotient of an even number of negatives is positive.

❑ The product or quotient of an odd number of negatives is negative.

Example Problems

These problems show the answers and solutions.

1. Multiply $(-2) \times (6)$.

    answer: -12

    There is one negative sign, so the answer is negative.

    $$(-2) \times (6) = -12$$

2. Multiply $(4) \times (-2) \times (-3) \times (7)$.

    answer: 168

    There are two negative signs, so the answer is positive.

    $$(4) \times (-2) \times (-3) \times (7) = 168$$

3. Divide $\dfrac{18}{-3}$.

    answer: -6

    There is one negative sign, so the answer is negative.

    $$\frac{18}{-3} = -6$$

4. Divide $(-28) \div (-4)$.

    answer: $+7$

    The are two negative signs, so the answer is positive.

    $$(-28) \div (-4) = +7$$

Work Problems

Use these problems to give yourself additional practice.

1. Multiply $-3 \times 4 \times 5$.

2. Multiply $(7) \times (-6) \times (-3)$.

3. Multiply $(-3) \times (-12) \times (-5)$.

4. Divide $-36 \div 4$.

5. Divide $\dfrac{-24}{-6}$.

Worked Solutions

1. **–60**   There is one negative sign, so the answer is negative.

2. **+126**   There are two negative signs, so the answer is positive.

3. **–180**   There are three negative signs, so the answer is negative.

4. **–9**   There is one negative sign, so the answer is positive.

5. **+4**   There are two negative signs, so the answer is positive.

Absolute Value

On a number line, the absolute value of a number tells how far that number is from 0. That means that the absolute value of 5 is 5, and the absolute value of -5 is 5 because both of these numbers are five units away from the origin.

The absolute value of a number is written by putting vertical bars around the number, like $|-5|$. You find the absolute value of a number by taking the number inside the absolute value signs and making it positive. Therefore, the absolute value of any number must always be positive. Remember, always carry out the operations inside the absolute value signs before taking the absolute value and then finish the problem.

Example Problems

These problems show the answers and solutions.

1. Evalute $|-6|$.

    answer: 6

    We make the number positive.

2. Evaluate $|9 - 12|$.

    answer: 3

    We simplify inside the vertical bars and then make the number positive:

    $|9 - 12| = |-3| = 3$

    Notice that this isn't the same as just making the sign inside the absolute value bars a plus:

    $9 + 12 = 21 \neq 3$

3. Evaluate $|+14|$.

    answer: 14

    The number inside is already positive, so you leave it as $+14$.

4. Evaluate $6 - |-24|$.

    answer: -18

    We make the number inside the vertical bars positive and then subtract:

    $6 - |-24| = 6 - 24 = -18$

Work Problems

Use these problems to give yourself additional practice.

1. Find the absolute value $|4|$.

2. Find the absolute value $|12 - 14|$.

3. Simplify $72 - |-64|$.

4. Simplify $|-200|$.

5. Simplify $|7| - |-13|$.

Worked Solutions

1. **4**   The number inside the vertical bars is already positive.

2. **2**   Subtract inside the vertical bars and then make the number positive:

    $$|12 - 14| = |-2| = 2$$

3. **8**   Make the number inside the vertical bars positive and then subtract:

$$72 - |-64| = 72 - 64 = 8$$

4. **200**   Make the number inside the vertical bars positive.

5. **–6**   Make the numbers inside the vertical bars positive and then subtract:

$$|7| - |-13| = 7 - 13 = -6$$

Notice that this is not the same as $|7 - -13| = |7 + 13| = |20| = 20$.

Rational Numbers

The group of numbers that consists of all the integers and all the fractions that have an integer for their numerators and an integer (excluding zero) for their denominators are called the rational numbers. Like the integers, the rationals include both negative and positive numbers and can be placed on a number line.

Negative Fractions

Fractions can be negative as well as positive. (See Chapter 3 to learn more about fractions.) Usually, negative fractions are written like $-\frac{2}{5}$ or $\frac{-2}{5}$ rather than $\frac{2}{-5}$, although all three fractions are exactly equal.

Adding Positive and Negative Fractions

The rules for what sign to keep when adding positive and negative fractions are the same as the rules for adding integers. Remember to get common denominators before adding fractions.

Example Problems

These problems show the answers and solutions.

1. Add $\frac{3}{7} + \frac{-2}{7}$.

   answer: $\frac{1}{7}$

   The signs are different; the answer is positive because $\frac{3}{7}$ is larger:

   $$\frac{3}{7} + \frac{-2}{7} = \frac{3 + -2}{7} = \frac{1}{7}$$

2. Add $\dfrac{-1}{3} + \dfrac{-3}{5}$.

    answer: $\dfrac{-14}{15}$

    The signs are the same, so the answer has that sign:

    $$\frac{-1}{3} + \frac{-3}{5} = \frac{-5}{15} + \frac{-9}{15} = \frac{-5 + -9}{15} = \frac{-14}{15}$$

3. Add $\dfrac{-1}{8} + \dfrac{3}{8}$.

    answer: $\dfrac{1}{4}$

    The signs are different, and the positive number is larger, so the answer is positive:

    $$\frac{-1}{8} + \frac{3}{8} = \frac{-1 + 3}{8} = \frac{\cancel{2}}{\cancel{8}} = \frac{1}{4}$$

4. Add $\dfrac{-4}{9} + \dfrac{1}{6}$.

    answer: $\dfrac{-5}{18}$

    The signs are different, and the absolute value of the negative number is larger, so the answer is negative:

    $$\frac{-4}{9} + \frac{1}{6} = \frac{-8}{18} + \frac{3}{18} = \frac{-8 + 3}{18} = \frac{-5}{18}$$

Work Problems

Use these problems to give yourself additional practice.

1. Add $\dfrac{-4}{11} + \dfrac{5}{11}$.

2. Add $\dfrac{-1}{2} + \dfrac{-1}{4}$.

3. Add $\dfrac{3}{5} + \dfrac{-4}{7}$.

4. Add $\dfrac{-1}{4} + \dfrac{7}{12}$.

5. Add $\dfrac{-2}{3} + \dfrac{-1}{2}$.

Worked Solutions

1. $\dfrac{1}{11}$ The positive number is larger, so the answer is positive:

$$\frac{-4}{11} + \frac{5}{11} = \frac{-4 + 5}{11} = \frac{1}{11}$$

2. $\dfrac{-3}{4}$ Both numbers are negative, so the answer is also negative:

$$\dfrac{-1}{2} + \dfrac{-1}{4} = \dfrac{-2}{4} + \dfrac{-1}{4} = \dfrac{-2 + -1}{4} = \dfrac{-3}{4}$$

3. $\dfrac{1}{35}$ The positive number is larger, so the answer is positive (notice that sometimes it's not clear which is larger until you have a common denominator!):

$$\dfrac{3}{5} + \dfrac{-4}{7} = \dfrac{21}{35} + \dfrac{-20}{35} = \dfrac{21 - 20}{35} = \dfrac{1}{35}$$

4. $\dfrac{1}{3}$ The positive number is larger, so the answer is positive:

$$\dfrac{-1}{4} + \dfrac{7}{12} = \dfrac{-3}{12} + \dfrac{7}{12} = \dfrac{-3 + 7}{12} = \dfrac{\cancel{4}}{\cancel{12}_3} = \dfrac{1}{3}$$

5. $\dfrac{-7}{6}$ Both numbers are negative, so the answer is negative:

$$\dfrac{-2}{3} + \dfrac{-1}{2} = \dfrac{-4}{6} + \dfrac{-3}{6} = \dfrac{-7}{6}$$

Adding Positive and Negative Mixed Numbers

The rules for adding positive and negative mixed numbers are the same as the rules you have been following when adding integers and rationals.

Example Problems

These problems show the answers and solutions.

1. Add $\left(-3\dfrac{1}{3}\right) + \left(-2\dfrac{1}{4}\right)$.

    answer: $-5\dfrac{7}{12}$

    Both numbers are negative, so the answer is negative:

    $$
    \begin{aligned}
    -3\tfrac{4}{12}& \\
    +\ -2\tfrac{3}{12}& \\
    \hline
    -5\tfrac{7}{12}&
    \end{aligned}
    $$

2. Add $-4\dfrac{7}{9} + 3\dfrac{5}{6}$.

    answer: $\dfrac{-17}{18}$

    Because the absolute value of the negative number is bigger, you subtract and make the answer negative:

$$-\begin{pmatrix} \overset{3}{\cancel{4}}\dfrac{14}{18} \\ -3\dfrac{15}{18} \end{pmatrix}$$

$$-\left(\dfrac{17}{18}\right)$$

Work Problems

Use these problems to give yourself additional practice.

1. Add $-4\dfrac{1}{5} + 6\dfrac{2}{5}$.

2. Add $-7\dfrac{5}{12} + 3\dfrac{4}{9}$.

3. Add $-2\dfrac{3}{4} + (-1\dfrac{1}{2})$.

4. Add $-5\dfrac{1}{5} + 2\dfrac{1}{3}$.

Worked Solutions

1. $2\dfrac{1}{5}$ The positive number is larger, so subtract the smaller and make the answer positive:

$$6\dfrac{2}{5}$$
$$-4\dfrac{1}{5}$$
$$\overline{2\dfrac{1}{5}}$$

2. $-3\dfrac{35}{36}$ The negative number is larger, so subtract the smaller and make the answer negative:

$$-\begin{pmatrix} \overset{6}{\cancel{7}}\overset{51}{\cancel{\dfrac{15}{36}}} \\ -3\,\dfrac{16}{36} \end{pmatrix}$$

$$-\left(3\,\dfrac{35}{36}\right)$$

3. $-4\dfrac{1}{4}$ Both numbers are negative, so add and make the answer negative:

$$-2\dfrac{3}{4} + (-1\dfrac{1}{2}) = -2\dfrac{3}{4} - 1\dfrac{2}{4} = -3\dfrac{5}{4} = -4\dfrac{1}{4}$$

4. $-2\dfrac{13}{15}$ The negative number is larger, so subtract and make the answer negative:

$$-\begin{pmatrix} \overset{4}{\cancel{5}}\overset{18}{\cancel{\dfrac{3}{15}}} \\ -2\,\dfrac{5}{15} \end{pmatrix}$$

$$-\left(2\,\dfrac{13}{15}\right)$$

Subtracting Positive and Negative Fractions

The steps needed to subtract positive and negative fractions are the same as they are for subtracting integers. Again, remember to get common denominators first.

Example Problems

These problems show the answers and solutions.

1. Subtract $\frac{3}{5} - \frac{4}{5}$.

   answer: $\frac{-1}{5}$

   Rewrite subtraction as adding the opposite. The absolute value of the negative number is bigger, so the answer is negative:

   $$\frac{3}{5} - \frac{4}{5} = \frac{3}{5} + \frac{-4}{5} = \frac{3 + -4}{5} = \frac{-1}{5}$$

2. Subtract $-\frac{2}{7} - \frac{1}{6}$.

   answer: $\frac{-19}{42}$ Rewrite subtraction as adding the opposite. Both numbers are negative, so the answer is negative:

   $$-\frac{2}{7} - \frac{1}{6} = -\frac{12}{42} - \frac{7}{42} = \frac{-19}{42}$$

3. Subtract $\frac{-4}{9} - \frac{-1}{2}$.

   answer: $\frac{1}{18}$

   Rewrite subtraction as adding the opposite. After you get a common denominator, you can see that the positive number is larger, so the answer is positive:

   $$\frac{-4}{9} - \frac{-1}{2} = \frac{-4}{9} + \frac{1}{2} = \frac{-8}{18} + \frac{9}{18} = \frac{-8 + 9}{18} = \frac{1}{18}$$

Work Problems

Use these problems to give yourself additional practice.

1. Subtract $\frac{1}{4} - \frac{3}{4}$.

2. Subtract $\frac{3}{10} - \frac{2}{3}$.

3. Subtract $\frac{-5}{9} - \frac{1}{3}$.

4. Subtract $\frac{-2}{5} - \frac{-3}{4}$.

5. Subtract $\frac{-2}{3} - \frac{3}{7}$.

Worked Solutions

1. $-\dfrac{1}{2}$ $\dfrac{1}{4} - \dfrac{3}{4} = \dfrac{1}{4} + \dfrac{-3}{4} = \dfrac{1-3}{4} = \dfrac{-2}{4} = \dfrac{-1}{2}$

2. $-\dfrac{11}{30}$ $\dfrac{3}{10} - \dfrac{2}{3} = \dfrac{3}{10} + \dfrac{-2}{3} = \dfrac{9}{30} + \dfrac{-20}{30} = \dfrac{9-20}{30} = \dfrac{-11}{30}$

3. $-\dfrac{8}{9}$ $\dfrac{-5}{9} - \dfrac{1}{3} = \dfrac{-5}{9} + \dfrac{-1}{3} = \dfrac{-5}{9} + \dfrac{-3}{9} = \dfrac{-5+-3}{9} = \dfrac{-8}{9}$

4. $\dfrac{7}{20}$ $\dfrac{-2}{5} - \dfrac{-3}{4} = \dfrac{-2}{5} + \dfrac{3}{4} = \dfrac{-8}{20} + \dfrac{15}{20} = \dfrac{-8+15}{20} = \dfrac{7}{20}$

5. $-\dfrac{23}{21}$ $\dfrac{-2}{3} - \dfrac{3}{7} = \dfrac{-2}{3} + \dfrac{-3}{7} = \dfrac{-14}{21} + \dfrac{-9}{21} = \dfrac{-14+-9}{21} = \dfrac{-23}{21}$

Note: Improper fractions and mixed numbers are interchangeable, so we provide answers in whichever form is most convenient. To convert from an improper fraction to a mixed number see Chapter 3.

Subtracting Positive and Negative Mixed Numbers

The same rules apply when subtracting positive and negative mixed numbers as apply when subtracting integers. Remember to find a common denominator, and if you have to borrow when subtracting, be especially careful.

Example Problems

These problems show the answers and solutions.

1. Subtract $2\dfrac{1}{4} - 3\dfrac{1}{2}$.

    answer: $-1\dfrac{1}{4}$

    Rewrite the subtraction as adding the opposite, $2\dfrac{1}{4} + -3\dfrac{1}{2}$, and then because the absolute value of the negative number is larger, the answer will be negative:

    $$-\left(\begin{array}{r} 3\dfrac{2}{4} \\ -2\dfrac{1}{4} \\ \hline \end{array} \right)$$
    $$-\left(1\dfrac{1}{4} \right)$$

2. Subtract $-4\dfrac{6}{7} - 3\dfrac{3}{4}$.

    answer: $-8\dfrac{17}{28}$

    $-4\dfrac{6}{7} - 3\dfrac{3}{4} = -4\dfrac{6}{7} + -3\dfrac{3}{4} = -4\dfrac{24}{28} + -3\dfrac{21}{28} = -7\dfrac{45}{28} = -8\dfrac{17}{28}$

Work Problems

Use these problems to give yourself additional practice.

1. Subtract $5\frac{1}{3} - 2\frac{2}{3}$.

2. Subtract $2\frac{6}{7} - 4\frac{1}{4}$.

3. Subtract $-3\frac{3}{5} - 1\frac{1}{3}$.

4. Subtract $4\frac{5}{6} - 7\frac{7}{8}$.

Worked Solutions

1. $2\frac{2}{3}$ $\begin{array}{r} \overset{4}{\cancel{5}}\frac{\overset{4}{1}}{3} \\ -2\frac{2}{3} \\ \hline 2\frac{2}{3} \end{array}$

2. $-1\frac{11}{28}$ $\begin{array}{r} -\left(\begin{array}{r} \overset{3}{\cancel{4}}\frac{\overset{35}{7}}{28} \\ -2\frac{24}{28} \\ \hline \end{array}\right) \\ -\left(1\frac{11}{28}\right) \end{array}$

3. $-4\frac{14}{15}$ $-3\frac{3}{5} - 1\frac{1}{3} = -3\frac{3}{5} + -1\frac{1}{3} = -3\frac{9}{15} + -1\frac{5}{15} = -4\frac{14}{15}$

4. $-3\frac{1}{24}$ $\begin{array}{r} -\left(\begin{array}{r} 7\frac{21}{24} \\ -4\frac{20}{24} \\ \hline \end{array}\right) \\ -\left(3\frac{1}{24}\right) \end{array}$

Multiplying Positive and Negative Fractions

The sign rules for multiplying integers are the same rules you follow when multiplying fractions. When multiplying fractions, multiply the two numerators, then the two denominators, and reduce if possible.

Example Problems

These problems show the answers and solutions.

1. Multiply $\frac{-2}{7} \cdot \frac{3}{5}$.

   answer: $\frac{-6}{35}$

   $$\frac{-2}{7} \cdot \frac{3}{5} = \frac{-2 \cdot 3}{7 \cdot 5} = \frac{-6}{35}$$

2. Multiply $\frac{-3}{4} \cdot \frac{-9}{10}$.

   answer: $\frac{27}{40}$

   $$\frac{-3}{4} \cdot \frac{-9}{10} = \frac{-3 \cdot -9}{4 \cdot 10} = \frac{27}{40}$$

3. Multiply $\frac{5}{6} \cdot \frac{-3}{7}$.

   answer: $\frac{-5}{14}$

   $$\frac{5}{6} \cdot \frac{-3}{7} = \frac{5 \cdot -3}{6 \cdot 7} = \frac{-15}{42} = \frac{-5}{14}$$

Work Problems

Use these problems to give yourself additional practice.

1. Multiply $\frac{2}{3} \times \frac{-4}{9}$.

2. Multiply $\frac{-6}{7} \times \frac{-5}{11}$.

3. Multiply $\frac{-2}{5} \cdot \frac{3}{4}$.

4. Multiply $\frac{+8}{13} \times \frac{+1}{2}$.

5. Multiply $\frac{-5}{8} \cdot \frac{-7}{8}$.

Worked Solutions

1. $\frac{-8}{27}$ $\frac{2}{3} \cdot \frac{-4}{9} = \frac{2 \cdot -4}{3 \cdot 9} = \frac{-8}{27}$

2. $\frac{30}{77}$ $\frac{-6}{7} \cdot \frac{-5}{11} = \frac{-6 \cdot -5}{7 \cdot 11} = \frac{30}{77}$

3. $\frac{-3}{10}$ $\frac{-2}{5} \cdot \frac{3}{4} = \frac{-2 \cdot 3}{5 \cdot 4} = \frac{-\overset{3}{\cancel{6}}}{\underset{10}{\cancel{20}}} = \frac{-3}{10}$

4. $\frac{4}{13}$ $\frac{+\overset{4}{\cancel{8}}}{13} \cdot \frac{+1}{\cancel{2}} = \frac{4}{13}$

5. $\frac{35}{64}$ $\frac{-5}{8} \cdot \frac{-7}{8} = \frac{-5 \cdot -7}{8 \cdot 8} = \frac{35}{64}$

Multiplying Positive and Negative Mixed Numbers

Again, the sign rules for integers apply for positive and negative mixed numbers. Remember to change the mixed number to an improper fraction before multiplying.

Example Problems

These problems show the answers and solutions.

1. Multiply $-2\frac{1}{2} \times 3\frac{4}{7}$.

   answer: $\frac{-125}{14}$

   $-2\frac{1}{2} \times 3\frac{4}{7} = \frac{-5}{2} \cdot \frac{25}{7} = \frac{-125}{14}$

2. Multiply $6\frac{1}{3} \times -4\frac{7}{8}$.

   answer: $\frac{-247}{8}$

   $6\frac{1}{3} \times -4\frac{7}{8} = \frac{19}{\cancel{3}} \cdot \frac{-\overset{13}{\cancel{39}}}{8} = \frac{-247}{8}$

Work Problems

1. Multiply $1\frac{3}{4} \times -4\frac{2}{5}$.

2. Multiply $-3\frac{4}{7} \cdot -5\frac{1}{6}$.

3. Multiply $-4\frac{3}{4} \times 1\frac{5}{6}$.

4. Multiply $2\frac{2}{5} \cdot -3\frac{1}{3}$.

Worked Solutions

1. $\frac{-77}{10}$ $1\frac{3}{4} \times -4\frac{2}{5} = \frac{7}{\underset{2}{\cancel{4}}} \cdot \frac{-\overset{11}{\cancel{22}}}{5} = \frac{-77}{10}$

2. $\frac{775}{42}$ $-3\frac{4}{7} - 5\frac{1}{6} = \frac{-25}{7} \cdot \frac{-31}{6} = \frac{-25 \cdot -31}{7 \cdot 6} = \frac{775}{42}$

3. $\frac{-209}{24}$ $-4\frac{3}{4} \times 1\frac{5}{6} = \frac{-19}{4} \cdot \frac{11}{6} = \frac{-19 \cdot 11}{4 \cdot 6} = \frac{-209}{24}$

4. -8 $2\frac{2}{5} - 3\frac{1}{3} = \frac{\overset{4}{\cancel{12}}}{\cancel{5}} \cdot \frac{-\overset{2}{\cancel{10}}}{\cancel{3}} = \frac{-8}{1} = -8$

Dividing Positive and Negative Fractions

The sign rules for dividing positive and negative fractions are exactly the same as the sign rules for dividing integers. To divide fractions, simply multiply the first number by the reciprocal of the second number.

Example Problems

These problems show the answers and solutions.

1. Divide $\dfrac{-3}{4} \div \dfrac{2}{5}$.

   answer: $\dfrac{-15}{8}$

   $$\dfrac{-3}{4} \div \dfrac{2}{5} = \dfrac{-3}{4} \cdot \dfrac{5}{2} = \dfrac{-15}{8}$$

2. Divide $\dfrac{6}{11} \div \dfrac{-3}{7}$.

   answer: $\dfrac{-14}{11}$

   $$\dfrac{6}{11} \div \dfrac{-3}{7} = \dfrac{6}{11} \cdot \dfrac{-7}{3} = \dfrac{\overset{2}{\cancel{6}}}{11} \cdot \dfrac{-7}{\cancel{3}} = \dfrac{-14}{11}$$

Work Problems

Use these problems to give yourself additional practice.

1. Divide $\dfrac{1}{3} \div \dfrac{-5}{6}$.

2. Divide $\dfrac{-5}{7} \div \dfrac{-4}{5}$.

3. Divide $\dfrac{-4}{9} \div \dfrac{2}{11}$.

4. Divide $\dfrac{-2}{3} \div \dfrac{1}{4}$.

5. Divide $\dfrac{-7}{9} \div \dfrac{-7}{12}$.

Worked Solutions

1. $\dfrac{-2}{5}$ $\dfrac{1}{3} \div \dfrac{-5}{6} = \dfrac{1}{3} \cdot \dfrac{-6}{5} = \dfrac{1}{\cancel{3}} \cdot \dfrac{-\overset{2}{\cancel{6}}}{5} = \dfrac{-2}{5}$

2. $\dfrac{25}{28}$ $\dfrac{-5}{7} \div \dfrac{-4}{5} = \dfrac{-5}{7} \cdot \dfrac{-5}{4} = \dfrac{25}{28}$

3. $\dfrac{-22}{9}$ $\dfrac{-4}{9} \div \dfrac{2}{11} = \dfrac{-\overset{2}{\cancel{4}}}{9} \cdot \dfrac{11}{\cancel{2}} = \dfrac{-22}{9}$

4. $\dfrac{-8}{3}$ $\dfrac{-2}{3} \div \dfrac{1}{4} = \dfrac{-2}{3} \cdot \dfrac{4}{1} = \dfrac{-8}{3}$

5. $\dfrac{4}{3}$ $\dfrac{-7}{9} \div \dfrac{-7}{12} = \dfrac{-7}{\underset{3}{\cancel{9}}} \cdot \dfrac{-\overset{4}{\cancel{12}}}{7} = \dfrac{4}{3}$

Dividing Positive and Negative Mixed Numbers

To divide positive and negative mixed numbers, use the sign rules used when dividing integers. First, change the mixed numbers into improper fractions. Then, take the reciprocal of the divisor and multiply.

Example Problems

These problems show the answers and solutions.

1. Divide $-3\frac{1}{4} \div \frac{2}{3}$.

    answer: $\frac{-39}{8}$

    $$-3\frac{1}{4} \div \frac{2}{3} = \frac{-13}{4} \div \frac{2}{3} = \frac{-13}{4} \cdot \frac{3}{2} = \frac{-39}{8}$$

2. Simplify $\dfrac{2\frac{2}{5}}{-4\frac{1}{8}}$.

    answer: $\frac{-32}{55}$

    $$\frac{2\frac{2}{5}}{-4\frac{1}{8}} = 2\frac{2}{5} \div -4\frac{1}{8} = \frac{12}{5} \div \frac{-33}{8} = \frac{\overset{4}{\cancel{12}}}{5} \cdot \frac{-8}{\underset{11}{\cancel{33}}} = \frac{-32}{55}$$

3. Simplify $\dfrac{-5\frac{1}{6}}{-2\frac{3}{4}}$.

    answer: $\frac{62}{33}$

    $$\frac{-5\frac{1}{6}}{-2\frac{3}{4}} = \frac{-31}{6} \div \frac{-11}{4} = \frac{-31}{\underset{3}{\cancel{6}}} \cdot \frac{\overset{2}{-\cancel{4}}}{11} = \frac{62}{33}$$

Work Problems

Use these problems to give yourself additional practice.

1. Divide $6\frac{2}{3} \div \frac{-2}{5}$.

2. Divide $-2\frac{4}{5} \div -1\frac{6}{7}$.

3. Simplify $\dfrac{\frac{-4}{9}}{2\frac{1}{2}}$.

4. Simplify $\dfrac{-3\frac{5}{7}}{-4\frac{1}{3}}$.

Worked Solutions

1. $\dfrac{-50}{3}$ $6\dfrac{2}{3} \div \dfrac{-2}{5} = \dfrac{\overset{10}{\cancel{20}}}{3} \cdot \dfrac{-5}{\cancel{2}} = \dfrac{-50}{3}$

2. $\dfrac{98}{65}$ $-2\dfrac{4}{5} \div -1\dfrac{6}{7} = \dfrac{-14}{5} \div \dfrac{-13}{7} = \dfrac{-14}{5} \cdot \dfrac{-7}{13} = \dfrac{98}{65}$

3. $\dfrac{-8}{45}$ $\dfrac{\dfrac{-4}{9}}{2\dfrac{1}{2}} = \dfrac{-4}{9} \div \dfrac{5}{2} = \dfrac{-4}{9} \cdot \dfrac{2}{5} = \dfrac{-8}{45}$

4. $\dfrac{6}{7}$ $\dfrac{-3\dfrac{5}{7}}{-4\dfrac{1}{3}} = \dfrac{-26}{7} \div \dfrac{-13}{3} = \dfrac{-\overset{2}{\cancel{26}}}{7} \cdot \dfrac{-3}{\cancel{13}} = \dfrac{6}{7}$

Chapter 7
Powers, Exponents, and Roots

Powers and Exponents

Powers and exponents are additional operations that go beyond the addition, subtraction, multiplication, and division that have been addressed so far.

Exponents

An **exponent** (also called a **power**) is a number that is placed to the right of and above another number. In the expression 3^4, the 4 is called the exponent, and the 3 is called the **base**. An exponent indicates the number of times to multiply the base number by itself, so 3^4 means a total of four threes multiplied, or $3^4 = 3 \times 3 \times 3 \times 3 = 81$. Read this expression as, "Three to the fourth power."

Any number with an exponent of 1 is just the number itself, so $x^1 = x$, and any nonzero number with an exponent of 0 is 1, so $x^0 = 1$ when x is any number except zero. These rules might seem strange at first, but they make operations with exponents work very smoothly, as you will see later.

Example Problems

These problems show the answers and solutions.

1. Expand 2^4.

    answer: 16

    $2^4 = 2 \times 2 \times 2 \times 2 = 16$

2. Expand 6^0.

    answer: 1

    $6^0 = 1$

3. Expand 6^3.

   answer: 216

   $6^3 = 6 \times 6 \times 6 = 216$

Work Problems

Use these problems to give yourself additional practice.

1. Expand 4^2.

2. Expand 5^1.

3. Expand 3^3.

4. Expand 7^2.

5. Expand 100^0.

Worked Solutions

1. **16** $4^2 = 4 \times 4 = 16$

2. **5** Any number to the first power is just the number itself.

3. **27** $3^3 = 3 \times 3 \times 3 = 27$

4. **49** $7^2 = 7 \times 7 = 49$

5. **1** Any nonzero number to the 0th power is 1.

Negative Exponents

If the exponent is a negative number, it means the reciprocal of the corresponding positive exponent, so $5^{-2} = \dfrac{1}{5^2} = \dfrac{1}{25}$.

Example Problems

These problems show the answers and solutions.

1. Expand 6^{-2}.

   answer: $\dfrac{1}{36}$

   $6^{-2} = \dfrac{1}{6^2} = \dfrac{1}{36}$

2. Expand 3^{-3}.

answer: $\frac{1}{27}$

$$3^{-3} = \frac{1}{3^3} = \frac{1}{27}$$

Work Problems

Use these problems to give yourself additional practice.

1. Expand 2^{-1}.

2. Expand 5^{-4}.

3. Expand 4^{-2}.

4. Expand $\left(\frac{1}{3}\right)^{-2}$.

5. Expand $\left(\frac{3}{4}\right)^{-3}$.

Worked Solutions

1. $\frac{1}{2}$ $2^{-1} = \frac{1}{2^1} = \frac{1}{2}$

2. $\frac{1}{625}$ $5^{-4} = \frac{1}{5^4} = \frac{1}{5 \times 5 \times 5 \times 5} = \frac{1}{625}$

3. $\frac{1}{16}$ $4^{-2} = \frac{1}{4^2} = \frac{1}{16}$

4. 9 $\left(\frac{1}{3}\right)^{-2} = \frac{1}{\left(\frac{1}{3}\right)^2} = \frac{1}{\frac{1}{3} \cdot \frac{1}{3}} = \frac{1}{\frac{1}{9}} = 9$

5. $\frac{64}{27}$ $\left(\frac{3}{4}\right)^{-3} = \frac{1}{\left(\frac{3}{4}\right)^3} = \frac{1}{\frac{3}{4} \cdot \frac{3}{4} \cdot \frac{3}{4}} = \frac{1}{\frac{27}{64}} = \frac{64}{27}$

Squares and Cubes

Two common types of powers are squares and cubes. **Squaring** a number means to multiply the number by itself: $5^2 = 5 \times 5 = 25$. A number like 25 is called a **perfect square** because it is the square of a whole number.

When you **cube** a number, you raise it to the third power. To do this, simply multiply three of the numbers: $5^3 = 5 \times 5 \times 5 = 125$. A perfect cube is any whole number cubed.

Work Problems

Use these problems to give yourself additional practice.

1. What is the square of 3?

2. What is 8 squared?

3. Is 20 a perfect square?

4. What is 8 cubed?

5. Is 125 a perfect cube?

Worked Solutions

1. **9** $3^2 = 3 \times 3 = 9$

2. **64** $8^2 = 8 \times 8 = 64$

3. **No**   20 is more than 16, which is the square of 4, but less than 25, which is the square of the next natural number, 5.

4. **512** $8^3 = 8 \times 8 \times 8 = 512$

5. **Yes** $125 = 5^3$, so 125 is a perfect cube.

Operations with Powers and Exponents

Convenient rules exist for multiplying or dividing numbers with exponents as long as the bases are the same.

❑ When multiplying numbers with the same base, keep the same base and add the exponents.

❑ When dividing numbers with the same base, subtract the exponent of the second number (or denominator) from the exponent in the first number (or numerator).

Example Problems

These problems show the answers and solutions.

1. Multiply $3^2 \times 3^4$.

   answer: 3^6

   You can use the rule and add the exponents:

   $3^2 \times 3^4 = 3^{2+4} = 3^6$

   Or, you can write out the multiplication as follows:

   $3^2 \times 3^4 = (3 \times 3) \times (3 \times 3 \times 3 \times 3)$

   $= 3 \times 3 \times 3 \times 3 \times 3 \times 3$

   $= 3^6$

   Both ways lead to the same answer, and that's why the rule works.

2. Multiply $5^2 \times 5^4$.

   answer: 5^6

   Because the bases are the same, add the exponents:

   $$5^2 \times 5^4 = 5^{2+4} = 5^6$$

   You can also multiply this out as $5 \times 5 \times 5 \times 5 \times 5 \times 5 = 15,625$, but often it's better to keep it in the form 5^6. Which way is better in a particular situation depends on what you have to do next, and you should use your judgment.

3. Multiply $4^1 \times 4^7$.

   answer: 4^8

   Because the bases are the same, add the exponents:

   $$4^{1+7} = 4^8 \text{ or } 65,536$$

4. Divide $4^3 \div 4^1$.

   answer: 4^2

   Because the bases are the same, subtract the exponents:

   $$4^{3-1} = 4^2 \text{ or } 16$$

5. Divide $\dfrac{7^5}{7^2}$.

   answer: 7^3

   Because the bases are the same, subtract the exponents:

   $$\frac{7^5}{7^2} = 7^{5-2} = 7^3 \text{ or } 343$$

Work Problems

1. Multiply $3^2 \times 3^1$.

2. Multiply $10^3 \times 10^2$.

3. Divide $\dfrac{2^5}{2^3}$.

4. Divide $7^3 \div 7^0$.

5. Write $3^3 \times 3^5 \div 3^2$ with a single exponent.

Worked Solutions

1. **3^3**   Because the bases are the same, add the exponents:

    $3^{2+1} = 3^3$ or 27

2. **10^5**   Because the bases are the same, add the exponents:

    $10^{3+2} = 10^5$ or 100,000

3. **2^2**   Because the bases are the same, subtract the exponents:

    $2^{5-3} = 2^2$ or 4

4. **7^3**   Because the bases are the same, subtract the exponents:

    $7^{3-0} = 7^3$ or 343

5. **3^6**   Because the bases are the same, add the exponents when multiplying and then subtract the exponents when dividing:

    $3^{3+5} \div 3^2 = 3^8 \div 3^2 = 3^{8-2} = 3^6$

More Operations with Powers and Exponents

If the numbers you are multiplying or dividing do not have the same base number, no convenient rule exists—you first must simplify the numbers with exponents and then multiply or divide accordingly. Similarly, if you are adding or subtracting numbers with exponents, you must simplify before adding or subtracting.

There is a convenient rule for exponents on numbers that already have exponents:

> When you raise a number with an exponent to another power, multiply the two exponents to find the final exponent. Keep the original base.

Although it's easy to use, many people have trouble remembering this rule and keeping it straight from the rules for adding and multiplying. If you think about it as you work through Example Problem 3, it might help you remember why the rule works.

Example Problems

These problems show the answers and solutions.

1. Multiply $3^2 \times 2^4$.

    answer: 144

    Because the bases are not the same, there is no convenient rule, and you must multiply everything out:

    $(3 \times 3) \times (2 \times 2 \times 2 \times 2) = 9 \times 16 = 144$

2.   Add $4^2 + 7^2$.

answer: 65

No convenient rule exists for addition with exponents, so you have to work everything out:

$(4 \times 4) + (7 \times 7) = 16 + 49 = 65$

3.   Write $\left(2^3\right)^4$ with a single exponent.

answer: 2^{12}

There are two ways to do this. The longer way is to handle this just like any other number raised to the fourth power, so take 2^3 four times and multiply (by adding exponents):

$2^3 \times 2^3 \times 2^3 \times 2^3 = 2^{3+3+3+3}$

On the other hand, if you use the rule given previously and multiply the exponents:

$2^{3 \times 4} = 2^{12}$

Both ways get the same answer, of course, but notice that the second way is really just a shorter version of the first one. In the first, you added four copies of the original exponent of 3; in the second you multiplied the original exponent of 3 by 4, which is the same thing.

4.   Write $\left(4^7\right)^5$ with a single exponent.

answer: 4^{35}

Again, you can do it by multiplying five copies of 4^7 (adding the exponents):

$4^7 \times 4^7 \times 4^7 \times 4^7 \times 4^7 = 4^{7+7+7+7+7} = 4^{35}$

or by using the rule (multiplying the exponents):

$4^{7 \times 5} = 4^{35}$

Work Problems

Use these problems to give yourself additional practice.

1.   Multiply $2^2 \times 2^5$.

2.   Divide $4^9 \div 4^7$.

3.   Multiply $5^2 \times 2^3$.

4.   Subtract $3^3 - 4^2$.

5.   Write $\left(5^2\right)^4$ with a single exponent.

Worked Solutions

1. **2^7 or 128**   Because the bases are the same, add the exponents:

    $$2^{2+5} = 2^7 \text{ or } 128$$

2. **4^2 or 16**   Because the bases are the same, subtract the exponents:

    $$4^{9-7} = 4^2 \text{ or } 16$$

3. **200**   Because the bases are different, multiply everything out:

    $$(5 \times 5) \times (2 \times 2 \times 2) = 25 \times 8 = 200$$

4. **11**   There's no easy rule for subtraction, so you have to work everything out:

    $$3 \times 3 \times 3 - 4 \times 4 = 27 - 16 = 11$$

5. **5^8**   You can use the rule and multiply the exponents:

    $$5^{2 \times 4} = 5^8$$

Square Roots and Cube Roots

A number can be raised to a power, like $4^2 = 16$ or $2^3 = 8$. This process also can be turned around by asking, "What number squares to give 16?" or "What number would I cube to get 8?" This reverse question is what square and cube roots are about.

Square Roots

The **square root** of a number is the number that, when squared, gives you the original number. For example, the square root of 36 is 6 because $6^2 = 6 \times 6 = 36$. Write square roots with the **radical sign**, which looks like $\sqrt{}$, with the number whose square root you're looking for inside. For example, the square root of 9 would be written $\sqrt{9}$.

You might notice that there are actually two numbers that qualify as square roots for any given positive number. For instance, 2 is a square root of 4 because $2^2 = 2 \times 2 = 4$, and -2 also is a square root of 4 because $(-2)^2 = -2 \times -2 = 4$. However, unless told otherwise, you want the positive answer.

You might also notice that, because the square of any real number will be nonnegative, square roots of negative numbers seem like a problem. We won't deal with that problem in this book, but it's worth remembering that the square root of a negative number cannot be a real number.

Example Problems

These problems show the answers and solutions.

1. Find the square root of 144.

    answer: 12

You are being asked to find a number that squares to give you 144. Because $12^2 = 12 \times 12 = 144$, 12 is the square root of 144. Notice that although it's fairly easy to verify that 12 is the square root of 144, there isn't an easy way to come up with 12 in the first place except hunches or trial and error.

2. Find the square root of 400.

answer: 20

You want a number whose square is 400. Because 400 ends in zero, you should suspect a number ending in 0 might work. Let's try 20: $20^2 = 20 \times 20 = 400$. So, 20 is the square root of 400.

Work Problems

Use these problems to give yourself additional practice.

1. Is the square root of 16 equal to 4?

2. Is the square root of 81 equal to 9?

3. Find the square root of 121.

4. Find the square root of 169.

5. Find the square root of 900.

Worked Solutions

1. **Yes, it is.** Because $4^2 = 4 \times 4 = 16$, the square root of 16 is 4.

2. **Yes, it is.** Because $9^2 = 9 \times 9 = 81$, the square root of 81 is 9.

3. **11** You should expect a number slightly larger than 10, since $10^2 = 100$ is less than 121. If you try 11, you see that $11^2 = 11 \times 11 = 121$, so 11 is the square root of 121.

4. **13** You should expect a number slightly larger than 12 because from an earlier example you know that the square root of 144 is 12. If you try 13, you see that $13^2 = 13 \times 13 = 169$, so 13 is the square root of 169.

5. **30** As in an earlier example, you should expect the square root of 900 to be a number ending in 0 and a number bigger than 20. If you try 30, you see that $30^2 = 30 \times 30 = 900$, so 30 is the square root of 900.

Cube Roots

The **cube root** of a number is the number that, when cubed, gives you the original number. For instance, the cube root of 8 is 2, because $2^3 = 8$. The symbol for the cube root is a radical sign with an **index** of 3, like $\sqrt[3]{8}$. This system also extends to talk about fourth roots (so $\sqrt[4]{81} = 3$), fifth roots (so $\sqrt[5]{32} = 2$), and beyond, but we won't say any more about those in this book.

Note: The index of a square root is 2, but we usually don't bother writing $\sqrt[2]{9} = 3$; we just write $\sqrt{9} = 3$. The index of 2 is understood, if nothing else is written.

Example Problems

These problems show the answers and solutions.

1. Is $\sqrt[3]{64}$ equal to 4?

   answer: Yes, it is.

   Because $4^3 = 4 \times 4 \times 4 = 16 \times 4 = 64$, we say that 4 is the cube root of 64.

2. Is $\sqrt[3]{-27}$ equal to -3?

   answer: Yes, it is.

   Because $(-3)^3 = -3 \times -3 \times -3 = 9 \times -3 = -27$, we say that -3 is the cube root of -27. (Notice that negatives work very differently with cube roots than they did with square roots.)

Work Problems

1. Is $\sqrt[3]{8}$ equal to 3?

2. Is $\sqrt[3]{125}$ equal to 5?

3. Find $\sqrt[3]{216}$.

Worked Solutions

1. **No, it is not.**   Because $3^3 = 27 \neq 8$, 3 is not the cube root of 8.

2. **Yes, it is.**   Because $5^3 = 5 \times 5 \times 5 = 25 \times 5 = 125$, we say that 5 is the cube root of 125.

3. **6**   Because the cube root of 125 is 5, you should expect a number slightly larger than 5. If you try 6, you see that $6^3 = 6 \times 6 \times 6 = 36 \times 6 = 216$, so we say that 6 is the cube root of 216.

Simplifying Square Roots

Some square roots can be simplified, even when they don't have perfect square roots. When simplifying a square root, first factor the number under the radical sign into two numbers, with one of the numbers a perfect square (it's best if you can find the largest possible perfect square that can be factored out). You then can pull out the perfect square and multiply that root with the number left under the radical sign.

Notice that some square roots cannot be simplified further because they are already in simplest form.

Example Problems

These problems show the answers and solutions.

1. Simplify $\sqrt{27}$.

    answer: $3\sqrt{3}$

    $\sqrt{27} = \sqrt{9 \cdot 3} = \sqrt{9} \cdot \sqrt{3} = 3\sqrt{3}$

2. Simplify $\sqrt{20}$.

    answer: $2\sqrt{5}$

    $\sqrt{20} = \sqrt{4 \cdot 5} = \sqrt{4} \cdot \sqrt{5} = 2\sqrt{5}$

3. Simplify $\sqrt{48}$.

    answer: $4\sqrt{3}$

    $\sqrt{48} = \sqrt{16 \cdot 3} = \sqrt{16} \cdot \sqrt{3} = 4\sqrt{3}$

4. Simplify $\sqrt{15}$.

    answer: $\sqrt{15}$

    Because there are no perfect squares that can be factored out of 15 (except 1, which doesn't help), $\sqrt{15}$ is already in simplest form.

Work Problems

Use these problems to give yourself additional practice.

1. Simplify $\sqrt{50}$.

2. Simplify $\sqrt{8}$.

3. Simplify $\sqrt{28}$.

4. Simplify $\dfrac{\sqrt{18}}{3}$.

5. Simplify $\dfrac{\sqrt{300}}{\sqrt{12}}$.

Worked Solutions

1. $\sqrt{50} = \sqrt{25 \cdot 2} = \sqrt{25} \cdot \sqrt{2} = 5\sqrt{2}$

2. $\sqrt{8} = \sqrt{4 \cdot 2} = \sqrt{4} \cdot \sqrt{2} = 2\sqrt{2}$

3. $\sqrt{28} = \sqrt{4 \cdot 7} = \sqrt{4} \cdot \sqrt{7} = 2\sqrt{7}$

4. $\dfrac{\sqrt{18}}{3} = \dfrac{\sqrt{9 \cdot 2}}{3} = \dfrac{\sqrt{9} \cdot \sqrt{2}}{3} = \dfrac{3\sqrt{2}}{3} = \sqrt{2}$

5. $\dfrac{\sqrt{300}}{\sqrt{12}} = \dfrac{\sqrt{100\cdot 3}}{\sqrt{4\cdot 3}} = \dfrac{\sqrt{100}\cdot\sqrt{3}}{\sqrt{4}\cdot\sqrt{3}} = \dfrac{10\sqrt{3}}{2\sqrt{3}} = \dfrac{10}{2} = 5$

Chapter 8

Powers of Ten and Scientific Notation

Powers of Ten

It is important, and also convenient, to be able to work with powers of ten because our number system is based on them. The notation for powers of ten is as follows:

$$10^0 = 1$$
$$10^1 = 10$$
$$10^2 = 10 \times 10 = 100$$
$$10^3 = 10 \times 10 \times 10 = 1000$$

Notice that the positive exponent on the 10 and the number of zeros after the 1 always match.

$$10^{-1} = \frac{1}{10} = 0.1$$

$$10^{-2} = \frac{1}{10^2} = \frac{1}{100} = 0.01$$

$$10^{-3} = \frac{1}{10^3} = \frac{1}{1000} = 0.001$$

Notice that the negative exponent on the 10 and the number of places the decimal point has been moved to the left of the 1 always match.

Multiplying Powers of Ten

When multiplying powers of ten, add the exponents.

Example Problems

These problems show the answers and solutions.

1. Multiply 100×100 and write the answer as a power of ten.

   answer: 10^4

   $$100 \times 100 = 10^2 \times 10^2 = 10^{2+2} = 10^4$$

2. Multiply $1000 \times 10,000$ and write the answer as a power of ten.

   answer: 10^7

   $1000 \times 10,000 = 10^3 \times 10^4 = 10^{3+4} = 10^7$

3. Multiply 0.001×10 and write the answer as a power of ten.

   answer: 10^{-2}

   $0.001 \times 10 = 10^{-3} \times 10^1 = 10^{-3+1} = 10^{-2}$

Work Problems
Use these problems to give yourself additional practice.

1. Multiply $10 \times 10 \times 10$ and write the answer as a power of ten.

2. Multiply 1000×0.1 and write the answer as a power of ten.

3. Multiply $10,000 \times 0.0001$ and write the answer as a power of ten.

4. Multiply $1000 \times 100,000$ and write the answer as a power of ten.

Worked Solutions

1. 10^3 $10 \times 10 \times 10 = 10^1 \times 10^1 \times 10^1 = 10^{1+1+1} = 10^3$

2. 10^2 $1000 \times 0.1 = 10^3 \times 10^{-1} = 10^{3-1} = 10^2$

3. 10^0 $10,000 \times 0.0001 = 10^4 \times 10^{-4} = 10^{4-4} = 10^0$

4. 10^8 $1000 \times 100,000 = 10^3 \times 10^5 = 10^{3+5} = 10^8$

Dividing Powers of Ten
When dividing powers of ten, subtract the exponent of the second number (the divisor) from the exponent of the first number.

Example Problems
These problems show the answers and solutions.

1. Divide $1000 \div 10$ and write the answer as a power of ten.

   answer: 10^2

   $1000 \div 10 = 10^3 \div 10^1 = 10^{3-1} = 10^2$

2. Divide $\frac{100}{0.1}$ and write the answer as a power of ten.

   answer: 10^3

   $$\frac{100}{0.1} = \frac{10^2}{10^{-1}} = 10^{2--1} = 10^3$$

Work Problems
Use these problems to give yourself additional practice.

1. Divide $10{,}000 \div 1000$ and write the answer as a power of ten.

2. Divide $\frac{100}{100{,}000}$ and write the answer as a power of ten.

3. Divide $1000 \div 0.01$ and write the answer as a power of ten.

4. Divide $0.0001 / 0.01$ and write the answer as a power of ten.

Worked Solutions

1. **10^1** $\quad 10{,}000 \div 1000 = 10^4 \div 10^3 = 10^{4-3} = 10^1$

2. **10^{-3}** $\quad \frac{100}{100{,}000} = \frac{10^2}{10^5} = 10^{2-5} = 10^{-3}$

3. **10^5** $\quad 1000 \div 0.01 = 10^3 \div 10^{-2} = 10^{3-2} = 10^5$

4. **10^{-2}** $\quad 0.0001 / 0.01 = 10^4 \div 10^{-2} = 10^{-4--2} = 10^{-2}$

Scientific Notation
When you deal with numbers that are quite large or quite small, you can use scientific notation. Using **scientific notation** means that you write the number as a number between 1 and 10 multiplied by a power of ten.

Example Problems
These problems show the answers and solutions.

1. Write 27,000,000 in scientific notation.

   answer: 2.7×10^7

   Move the decimal point so that you have a number between 1 and 10 and then count the number of places you had to move it and use that number as the power of ten:

   $$27{,}000{,}000 = 2.7 \times 10^7$$

   Notice that the exponent is positive because the decimal point in 2.7 would need to be moved to the right seven places to recover 27,000,000.

2. Write 0.0000000094 in scientific notation.

    answer: 9.4×10^{-9}

    Write your number, which is between 1 and 10, and then count the number of places you moved the decimal point to get your power of ten:

    $0.0000000094 = 9.4 \times 10^{-9}$

    Notice that this time the exponent is negative because the decimal point in 9.4 would need to be moved 9 places to the left to recover 0.0000000094.

Remember when dealing with large whole numbers that your exponent will be positive, but when dealing with small decimals, your exponent will be negative.

Work Problems
Use these problems to give yourself additional practice.

1. Write 380,000 in scientific notation.

2. Write 0.00015 in scientific notation.

3. Write 42.9 in scientific notation.

4. Write 0.000067 in scientific notation.

Worked Solutions

1. **3.8×10^5** $380,000 = 3.8 \times 10^5$

2. **1.5×10^{-4}** $0.00015 = 1.5 \times 10^{-4}$

3. **4.29×10^1** $42.9 = 4.29 \times 10^1$

4. **6.7×10^{-5}** $0.000067 = 6.7 \times 10^{-5}$

Multiplying in Scientific Notation
When multiplying numbers in scientific notation, first multiply the numbers between 1 and 10 and then add up the exponents.

Example Problems
These problems show the answers and solutions.

1. Multiply $(2 \times 10^3) \times (3 \times 10^4)$ and give the answer in scientific notation.

    answer: 6×10^7

    Multiply the numbers and add the exponents.

2. Multiply $(3 \times 10^3) \times (4 \times 10^5)$ and give the answer in scientific notation.

    answer: 1.2×10^9

    Proceed as before and then notice that at this stage the number 12 is not between 1 and 10. You can fix this by moving the decimal point one place and adjusting the exponent on the 10:

    $$12 \times 10^8 = 1.2 \times 10^9$$

3. Multiply $(6 \times 10^5) \times (5 \times 10^{-3})$ and give the answer in scientific notation.

    answer: 3×10^3

    $(6 \times 10^5) \times (5 \times 10^{-3}) = 30 \times 10^2 = 3 \times 10^3$

Work Problems

Use these problems to give yourself additional practice.

1. Multiply $(4 \times 10^3) \times (2 \times 10^4)$ and give the answer in scientific notation.

2. Multiply $(5 \times 10^6) \times (3 \times 10^2)$ and give the answer in scientific notation.

3. Multiply $(3 \times 10^{-4}) \times (3 \times 10^5)$ and give the answer in scientific notation.

4. Multiply $(9 \times 10^{-3}) \times (4 \times 10^{-7})$ and give the answer in scientific notation.

Worked Solutions

1. $\mathbf{8 \times 10^7}$ Multiply the numbers and add the exponents.

    $$(4 \times 10^3) \times (2 \times 10^4) = 8 \times 10^7$$

2. $\mathbf{1.5 \times 10^9}$ Multiply the numbers and add the exponents; then adjust the decimal point.

    $$(5 \times 10^6) \times (3 \times 10^2) = 15 \times 10^8 = 1.5 \times 10^9$$

3. $\mathbf{9 \times 10^1}$ Multiply the numbers and add the exponents.

    $$(3 \times 10^{-4}) \times (3 \times 10^5) = 9 \times 10^1$$

4. $\mathbf{3.6 \times 10^{-9}}$ Multiply the numbers and add the exponents; then adjust the decimal point.

    $$(9 \times 10^{-3}) \times (4 \times 10^{-7}) = 36 \times 10^{-10} = 3.6 \times 10^{-9}$$

Dividing in Scientific Notation

When dividing two numbers in scientific notation, first divide the numbers that are between 1 and 10. Then, subtract the second exponent from the first exponent to get the new power of ten. As in multiplication, you may need to adjust your answer to ensure that it is still in scientific notation.

Example Problems

These problems show the answers and solutions.

1. Divide $(9 \times 10^3) \div (3 \times 10^1)$ and give the answer in scientific notation.

    answer: 3×10^2

    Divide the numbers and subtract the exponents

    $(9 \times 10^3) \div (3 \times 10^1) = 3 \times 10^2$.

2. Divide $\dfrac{2 \times 10^9}{8 \times 10^6}$

    and give the answer in scientific notation.

    answer: 2.5×10^2

    Divide the numbers and subtract the exponents and then adjust the decimal point:

    $$\frac{2 \times 10^9}{8 \times 10^6} = 0.25 \times 10^3 = 2.5 \times 10^2$$

Work Problems

Use these problems to give yourself additional practice.

1. Divide $(6 \times 10^4) \div (3 \times 10^3)$ and give the answer in scientific notation.

2. Divide $\dfrac{8 \times 10^{-3}}{2 \times 10^4}$ and give the answer in scientific notation.

3. Divide $(9 \times 10^{-5}) / (4 \times 10^{-8})$ and give the answer in scientific notation.

4. Divide $(6 \times 10^6) \div (8 \times 10^4)$ and give the answer in scientific notation.

Worked Solutions

1. 2×10^1 Divide the numbers and subtract the exponents:

    $$(6 \times 10^4) \div (3 \times 10^3) = 2 \times 10^1$$

2. 4×10^{-7} Divide the numbers and subtract the exponents:

    $$\frac{8 \times 10^{-3}}{2 \times 10^4} = 4 \times 10^{-7}$$

3. **2.25×10^3** Divide the numbers and subtract the exponents:

$$(9 \times 10^{-5}) / (4 \times 10^{-8}) = 2.25 \times 10^3$$

4. **7.5×10^1** Divide the numbers and subtract the exponents and then adjust the decimal point:

$$(6 \times 10^6) \div (8 \times 10^4) = 0.75 \times 10^2 = 7.5 \times 10^1$$

Other Operations in Scientific Notation

There are no special procedures for operations like addition or subtraction in scientific notation. If the powers of ten match, you can simply add or subtract the numbers and keep the same power of ten (you may, of course, have to adjust the decimal point). If the powers of ten do not match, it may be easiest to convert the numbers back to regular notation, perform the addition or subtraction, and then change that answer back to scientific notation.

To apply an exponent to a number in scientific notation, apply the exponent to the number between 1 and 10 and multiply the exponent times the original exponent on the 10. Adjust the decimal point if necessary.

Example Problems

These problems show the answers and solutions.

1. Add $3.2 \times 10^3 + 9.1 \times 10^3$ and give the answer in scientific notation.

   answer: 1.23×10^4

   This time you can stay in scientific notation and add:

   $3.2 \times 10^3 + 9.1 \times 10^3 = (3.2 + 9.1) \times 10^3 = 12.3 \times 10^3 = 1.23 \times 10^4$

   Or, you can convert to regular notation, add, and change back to scientific notation:

   $3.2 \times 10^3 + 9.1 \times 10^3 = 3200 + 9100 = 12300 = 1.23 \times 10^4$.

2. Subtract $4.3 \times 10^5 - 7.2 \times 10^4$ and give the answer in scientific notation.

   answer: 3.58×10^5

   Because the exponents do not match, the best thing to do is change to regular notation, subtract, and then change back to scientific notation:

   $4.3 \times 10^5 - 7.2 \times 10^4 = 430,000 - 72,000 = 358,000 = 3.58 \times 10^5$

3. Simplify $(2 \times 10^4)^3$ and give the answer in scientific notation.

   answer: 8×10^{12}

   Raise 2 to the third power and multiply the exponents:

   $$(2 \times 10^4)^3 = 2^3 \times 10^{4 \times 3} = 8 \times 10^{12}$$

Work Problems

Use these problems to give yourself additional practice.

1.   Add $7.5 \times 10^8 + 5.0 \times 10^7$ and give the answer in scientific notation.

2.   Subtract $4.6 \times 10^4 - 4.2 \times 10^4$ and give the answer in scientific notation.

3.   Simplify $(5 \times 10^4)^2$ and give the answer in scientific notation.

4.   Simplify $(6 \times 10^7)^3$ and give the answer in scientific notation.

Worked Solutions

1.   **8×10^8**   Change to regular notation, add, and change back to scientific notation:

$$7.5 \times 10^8 + 5.0 \times 10^7 = 750{,}000{,}000 + 50{,}000{,}000 = 800{,}000{,}000 = 8 \times 10^8$$

2.   **4×10^3**   Because the exponents are the same, you can subtract (and then adjust the decimal point):

$$4.6 \times 10^4 - 4.2 \times 10^4 = 0.4 \times 10^4 = 4 \times 10^3$$

3.   **2.5×10^9**   Raise 5 to the second power and multiply the exponents; then adjust the decimal point:

$$(5 \times 10^4)^2 = 5^2 \times 10^{4 \times 2} = 25 \times 10^8 = 2.5 \times 10^9$$

4.   **2.16×10^{23}**   Raise 6 to the third power and multiply the exponents; then adjust the decimal point:

$$(6 \times 10^7)^3 = 6^3 \times 10^{7 \times 3} = 216 \times 10^{21} = 2.16 \times 10^{23}$$

Chapter 9
Measurements

There are two commonly used systems of measurement: the English system and the metric system. We will deal with both here, as well as changing measurements from one system to the other. Also, at the end of the chapter we will cover the basics of finding perimeters and areas of some common shapes.

The English System

The **English system** of measurement is still in use in the United States for many nonscientific purposes. The many units of measure and conversion factors in the English system can be very difficult to remember and use.

The Units in the English System

Following is a summary of the most common units in the English system:

Length is usually measured in **inches** (abbreviated in), **feet** (abbreviated ft), and **miles** (abbreviated mi).

> 12 inches (in) = 1 foot (ft)
> 3 feet (ft) = 1 yard (yd)
> 5280 feet (ft) = 1 mile (mi)

Area is mostly measured in units derived from measures of length; for instance, a square patch of area that measures 1 inch on each side is said to have an area of 1 square inch. The commonly used units are **square inches** (abbreviated sq in or in^2), **square feet** (abbreviated sq ft or ft^2), **acres**, and **square miles** (abbreviated sq mi or mi^2).

> 144 square inches (in^2) = 1 square foot (ft^2)
> 9 square feet (ft^2) = 1 square yard (yd^2)
> 4840 square yards (yd^2) = 1 acre
> 640 acres = 1 square mile (mi^2)

Weight is typically measured in **ounces** (abbreviated oz), **pounds** (abbreviated lbs—notice that this abbreviation doesn't make much sense in modern English), and **tons**.

> 16 ounces (oz) = 1 pound (lb)
> 2000 pounds (lbs) = 1 ton

Capacity is typically measured in **fluid ounces** (abbreviated fl oz), **cups**, **quarts** (abbreviated qt), and **gallons** (abbreviated gal).

> 8 fluid ounces (fl oz) = 1 cup
> 4 cups = 1 quart (qt)
> 4 quarts (qt) = 1 gallon (gal)

Time is measured in **seconds** (abbreviated sec), **minutes** (abbreviated min), **hours** (abbreviated hr), **days**, **weeks**, **years**, **decades**, and **centuries**.

> 60 seconds (sec) = 1 minute (min)
> 60 minutes (min) = 1 hour (hr)
> 24 hours (hr) = 1 day
> 7 days = 1 week
> 365 days ≈ 1 year
> 10 years = 1 decade
> 100 years = 1 century

Changing Units within the English System

Often we want to know about equivalences that aren't given specifically in this summary—for instance, the number of hours in 3 days or maybe even the number of inches in half a mile. The easiest way to handle these questions is to recognize that we can rearrange, multiply, and combine any of the equations given previously.

Example Problems

These problems show the answers and solutions.

1. How many hours are there in 3 days?

    answer: 72 hours

    You know that 24 hours = 1 day. You can multiply both sides of this equation by 3 to get 3×24 hours = 3×1 day.

    $$72 \text{ hours} = 3 \text{ days}$$

2. How many inches are there in half a mile?

    answer: 31,680 inches

    You know that 12 inches = 1 foot and that 5280 feet = 1 mile, so if you multiply the first equality by 5280 and put these together, you have 5280×12 inches = 5280×1 foot = 1 mile, or 63360 inches = 1 mile.

Now you know a relationship between inches and miles. Because you wanted to know how many inches are in half a mile, you could divide both sides of the equation by 2:

$$\frac{63,360 \text{ inches}}{2} = \frac{1 \text{ mile}}{2}$$

$$31,680 \text{ inches} = \frac{1}{2} \text{ mile}.$$

3. How many fluid ounces are there in a gallon?

answer: 128 fluid ounces

You know that 8 fluid ounces = 1 cup, 4 cups = 1 quart, and 4 quarts = 1 gallon. You can multiply the first equality by 4 and combine it with the second to conclude that 32 fluid ounces = 1 quart, and then multiply this by 4 and combine it with the other equality to find that 128 fluid ounces = 1 gallon.

Work Problems

Use these problems to give yourself additional practice.

1. How many inches are there in 15 feet?

2. How many days are there in a decade?

3. How many cups are there in a gallon?

4. How many seconds are there in 4 hours?

5. How many ounces are there in a quarter ton?

Worked Solutions

1. **180 inches**   Since 12 inches = 1 foot, multiply both sides of this equation by 15 to get 15 × 12 inches = 15 × 1 foot, or 180 inches = 15 feet.

2. **3650 days**   Because there are 365 days ≈ 1 year and 10 years = 1 decade, multiply the first equation by 10 to get 10 × 365 days ≈ 10 × 1 year and combine this with the second equation to get 10 × 365 days ≈ 10 years = 1 decade, or just 3650 days ≈ 1 decade.

3. **16 cups**   You know that 4 cups = 1 quart and 4 quarts = 1 gallon, so if you multiply the first equation by 4 to get 4 × 4 cups = 4 × 1 quart and combine this with the second equation, then you have 4 × 4 cups = 1 gallon, or 16 cups = 1 gallon.

4. **14,400 seconds**   You know that 60 seconds = 1 minute and 60 minutes = 1 hour. If you multiply the first equation by 60, you have 360 seconds = 60 minutes; combining this with the second equation tells you that 360 seconds = 1 hour. Now, to find out about 4 hours, you can multiply both sides of this equation by 4 to get 4 × 360 seconds = 4 × 1 hour, or 14,400 seconds = 4 hours.

5. **8000 ounces**   You know that 16 ounces = 1 pound and 2000 pounds = 1 ton. If you multiply the first equation by 2000, you get 32,000 ounces = 2000 pounds, and combining this with the second equation tells you that 32,000 ounces = 1 ton. Now because you originally wanted to know about $\frac{1}{4}$ ton, divide both sides of this equation by 4 to get 8000 ounces = $\frac{1}{4}$ ton.

The Metric System

Most of the world outside the United States, and scientists everywhere, primarily use the **metric system** (sometimes abbreviated as the **SI system**). The metric system has a few basic units, and measures larger and smaller quantities using multiples of ten. The standard metric units include **meters** (abbreviated m) for length, **grams** (abbreviated g) for mass, and **liters** (abbreviated l) for volume. The metric system uses seconds, minutes, and hours for time just like the English system.

For larger and smaller amounts, the metric system uses the following prefixes:

kilo-	for thousand
hecto-	for hundred
deca-	for ten
deci-	for tenth
centi-	for hundredth
milli-	for thousandth

This means that one kilometer is a thousand meters, a milligram is one thousandth of a gram, and so forth. This makes changing between units in the metric system much easier than in the English system.

Example Problems

These problems show the answers and solutions.

1. How many centimeters are there in a kilometer?

    answer: 100,000 centimeters

    You know that 1 kilometer = 1000 meters, and 1 meter = 100 centimeters. If you multiply the second equation by 1000 and put it together with the first, you see that 1 kilometer = 100,000 centimeters.

2. How many grams are in 5 kilograms?

    answer: 5000 grams

    We know that 1000 grams = 1 kilogram, so we multiply by 5 to see that 5000 grams = 5 kilograms.

Work Problems

Use these problems to give yourself additional practice.

1. How many centimeters are there in a meter?

2. How many deciliters are there in a liter?

3. How many grams are there in a kilogram?

4. How many millimeters are there in a centimeter?

5. How many milliliters are there in a kiloliter?

Worked Solutions

1. **100 centimeters**   Since centi- means hundredth, there are 100 centimeters in a meter.

2. **10 deciliters**   Since deci- means tenth, there are 10 deciliters in a liter.

3. **1000 grams**   Since kilo- means thousand, there are 1000 grams in a kilogram.

4. **10 millimeters**   Since milli- means thousandth, 1000 millimeters = 1 meter. Since centi- means hundredth, 100 centimeters = 1 meter. Putting these together gives you 1000 millimeters = 100 centimeters, and dividing by 100 gives 10 millimeters = 1 centimeter.

5. **1,000,000 milliliters**   Since milli- means thousandth, 1000 milliliters = 1 liter. Since kilo- means thousand, 1000 liters = 1 kiloliter. If we multiply the first equation by 1000 to get 1,000,000 milliliters = 1000 liters and combine with the second equation, then we have 1,000,000 milliliters = 1 kiloliter.

Changing between English and Metric Units

Changing between English and metric units is inconvenient because the conversion factors involved are not whole numbers. However, it is often necessary to make such changes, partly because the United States is gradually shifting from one system to the other. The following table gives several useful equivalences:

	English to Metric	*Metric to English*
Length	1 inch ≈ 2.54 centimeters	1 centimeter ≈ 0.39 inches
	1 foot ≈ 30.48 centimeters	1 meter ≈ 39.37 inches
	1 mile ≈ 1.61 kilometers	1 kilometer ≈ 0.62 miles
Weight and Mass	1 ounce ≈ 28.35 grams	1 gram ≈ 0.035 ounces
	1 pound ≈ 0.45 kilograms	1 kilogram ≈ 2.20 pounds
Capacity	1 fluid ounce ≈ 29.57 milliliters	1 milliliter ≈ 0.034 fluid ounces
	1 gallon ≈ 3.78 liters	1 liter ≈ 0.26 gallons

Example Problems

These problems show the answers and solutions.

1. How many centimeters are there in 2 feet?

    answer: ≈ 60.96 centimeters

    Since you know that 1 foot ≈ 30.48 centimeters, you multiply both sides by 2 to get 2 feet ≈ 60.96 centimeters.

2. How many kilograms are in a ton?

    answer: ≈ 900 kilograms

    You know that 1 ton = 2000 pounds and 1 pound ≈ 0.45 kilograms. If you multiply the second equivalence by 2000 and combine with the first, you have 1 ton ≈ 900 kilograms.

3. How many fluid ounces are there in 2-liter bottle?

    answer: ≈ 66.56 fluid ounces

    You know that 1 liter ≈ 0.26 gallons, and from Example Problem 3 in "Changing Units within the English System," you know that 1 gallon = 128 fluid ounces. If you multiply both sides of this second equality by 0.26 and combine with the first, you can conclude that 1 liter ≈ 33.28 fluid ounces. Since you wanted to know in the first place about a 2-liter bottle, multiply both sides by 2 to find that 2 liters ≈ 66.56 fluid ounces.

Work Problems

Use these problems to give yourself additional practice.

1. If a man is 6 feet tall, how many meters tall is he?

2. If a beverage can contains 12 fluid ounces, how many milliliters does it contain?

3. If a piece of equipment is labeled as 3000 kilograms, should it be moved by a forklift rated for a maximum of 5000 pounds?

4. If a woman is 5 feet, 4 inches tall, how many meters tall is she?

5. If a car is traveling 65 miles per hour, how many kilometers per hour is it going?

Worked Solutions

1. **≈ 1.8288 meters**   If 1 foot ≈ 30.48 centimeters, then 6 feet ≈ 182.88 centimeters. Dividing by 100 to change centimeters to meters, you have 6 feet ≈ 1.8288 meters.

2. **≈ 354.84 milliliters**   Since 1 fluid ounce ≈ 29.57 milliliters, you multiply both sides by 12 to get 12 fluid ounces ≈ 354.84 milliliters.

3. **No, it would be too heavy.** We know that 1 kilogram ≈ 2.20 pounds, so we multiply both sides by 3000 to get 3000 kilograms ≈ 6600 pounds. Since this is more than 5000 pounds, it is probably not safe to lift the equipment with the forklift.

4. **≈ 1.6256 meters** Since an inch is one twelfth of a foot, 4 inches is $\frac{1}{3}$ feet, and thus the woman is $5\frac{1}{3}$ feet tall. You know that 1 foot ≈ 30.48 centimeters. Then you multiply both sides by $5\frac{1}{3}$ to get $5\frac{1}{3}$ feet ≈ 162.56 centimeters, or 1.6256 meters.

5. **≈ 104.65 kilometers per hour** You know that 1 mile ≈ 1.61 kilometers. Then if you multiply both sides by 65, you get 65 miles per hour ≈ 104.65 kilometers per hour.

Significant Digits

You may have noticed in the previous section that the ≈ symbol was used frequently. One further problem with approximate equivalences, and especially with measured quantities, is that rounding can lead to additional errors. The usual way this problem is addressed in the sciences is by counting **significant digits**, and then rounding final answers to the number of digits in whichever of the original values had the fewest significant digits.

Digits are generally considered significant if they are nonzero, or zeros between nonzero digits. If there are trailing zeros beyond the last nonzero digit to the right of the decimal point, these are considered significant as well.

Example Problems

These problems show the answers and solutions.

1. How many significant digits are there in 6500?

   answer: 2

   The trailing zeros don't count, so 6500 has two significant digits (the 6 and the 5).

2. How many significant digits are there in 30.48?

   answer: 4

   All the digits are significant, so 30.48 has four significant digits.

3. How many significant digits are there in 0.007?

   answer: 1

   The initial zeros are not significant, so there is only one significant digit (the 7) in 0.007.

4. How many significant digits are there in 0.00700?

   answer: 3

   The initial zeros are not significant, but the two trailing zeros are. So 0.00700 has three significant digits (700).

5. How many significant digits are there in 40.007?

   answer: 5

   Since all the zeros are between nonzero digits, every digit is significant. Thus, 40.007 has five significant digits.

6. How many fluid ounces (counting significant digits) are there in 2-liter bottle?

   answer: ≈ 67 fluid ounces

   Notice that this is the same question as in "Changing between English Metric Units," where you arrived at the conclusion that 2 liters ≈ 66.56 fluid ounces. In that process you used the approximation 1 liter ≈ 0.26 gallons, and the number on the right-hand side has only two significant digits (it might be argued that other values involved, such as the "1 liter" on the left-hand side, seem to have only one significant digit, but actually these are exact values and significant digits don't apply). So our answer of 66.56 fluid ounces should be rounded to two digits, or 2 liters ≈ 67 fluid ounces.

Work Problems

Use these problems to give yourself additional practice.

1. How many significant digits are there in 0.62?

2. How many significant digits are there in 93,000,000?

3. How many significant digits are there in 0.050?

4. If a beverage can contains exactly 12 fluid ounces, how many milliliters (counting significant digits) does it contain?

5. If a scale says a person weighs 182 pounds, how many kilograms (counting significant digits) is that person?

Worked Solutions

1. **2**   The leading zero is not significant, but the 6 and 2 are, so there are two significant digits in 0.62.

2. **2**   The trailing zeros are not significant, but the 9 and 3 are, so there are two significant digits in 93,000,000.

3. **2**   The leading zeros are not significant, but the 5 and the trailing zero (since it's to the right of the decimal place and the last nonzero digit) are, so there are two significant digits in 0.050.

4. **≈ 354.8 milliliters**   In Work Problem 2 in the preceding section you concluded that 12 fluid ounces ≈ 354.84 milliliters. Since you used a conversion factor of 1 fluid ounce ≈ 29.57 milliliters, where the 29.57 has four significant digits, you should round 354.84 milliliters to four significant digits, or 12 fluid ounces ≈ 354.8 milliliters.

5. ≈ **82 kilograms**   You know that 1 pound ≈ 0.45 kilograms, so multiplying both sides of this by 182 gives you 182 pounds ≈ 81.9 kilograms. Since 0.45 has only two significant digits, you round your answer to two digits and say 182 pounds ≈ 82 kilograms.

Measurement of Basic Figures

A **polygon** is a figure that can be drawn using only straight-line segments. We will cover the basics of finding the **perimeter**, which is the distance around the outside, and **area**, which is the measure of the region inside, of some simple polygons.

First, we should introduce some basic types of polygons. A **triangle** is a polygon with exactly three sides. If one of the angles is a **right angle**, which means a 90-degree angle, like one of the corners of a square, then the triangle is called a **right triangle**. An **isosceles triangle** has two sides of equal length, and an **equilateral triangle** has all three sides of equal lengths.

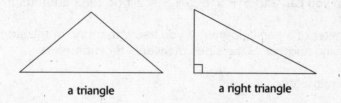

a triangle   a right triangle

A **quadrilateral** is a polygon with exactly four sides. A quadrilateral that has two sides parallel to each other is called a **trapezoid**. If both pairs of opposite sides are parallel to each other, the quadrilateral is called a **parallelogram**. It turns out that in a parallelogram the opposite sides will always be of equal lengths (the reasons for this fact are covered in geometry classes). A parallelogram with right angles is called a **rectangle**, and a rectangle with all four sides equal in length is called a **square**.

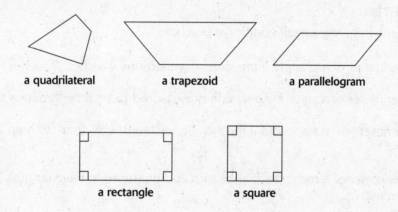

a quadrilateral   a trapezoid   a parallelogram

a rectangle   a square

Perimeter

Finding the perimeter of a triangle, quadrilateral, or other polygon is easy if the length of each side is given. Simply add up the lengths of all sides, and the total is the perimeter of the figure.

Example Problems

These problems show the answers and solutions.

1. Find the perimeter of a triangle with sides of lengths 3 inches, 4 inches, and 5 inches.

    answer: 12 inches

    You add the lengths of the three sides, $3 + 4 + 5 = 12$. So the perimeter of the triangle is 12 inches.

2. Find the perimeter of a square with one side of length 5 meters.

    answer: 20 meters

    Since all four sides of a square have the same lengths, you must have four sides of length 5 meters. Then you can add $5 + 5 + 5 + 5 = 20$, so the perimeter is 20 meters.

3. Find the perimeter of a parallelogram if you know that one of the sides measures 6 centimeters and another of the sides measures 8 centimeters.

    answer: 28 centimeters

    You know that a parallelogram must have opposite sides of equal length, so the third and fourth sides must have lengths 6 centimeters and 8 centimeters as well. Then you can add up these four lengths, $6 + 8 + 6 + 8 = 28$, so the perimeter of the parallelogram must be 28 centimeters.

Work Problems

Use these problems to give yourself additional practice.

1. Find the perimeter of a triangle if the sides have lengths 4 inches, 7 inches, and 6 inches.

2. Find the perimeter of a square if one side is measured to be 8 centimeters long.

3. Find the perimeter of a trapezoid if the sides have lengths 30 feet, 20 feet, 20 feet, and 10 feet.

4. Find the perimeter of a rectangle if two sides are measured to have lengths 20 centimeters and 50 centimeters.

5. Find the perimeter of an equilateral triangle if one side is measured to have a length of 4 meters.

Worked Solutions

1. **17 inches**   You add the side lengths, so $4 + 7 + 6 = 17$ inches is the perimeter.

2. **32 centimeters**   All the sides must be 8 centimeters long, so the perimeter is $8 + 8 + 8 + 8 = 32$ centimeters.

3. **80 feet**   You add the four sides, so 30 + 20 + 20 + 10 = 80 feet.

4. **140 centimeters**   The two other sides must also have lengths of 20 centimeters and 50 centimeters, so the perimeter is 20 + 50 + 20 + 50 = 140 centimeters.

5. **12 meters**   The sides of an equilateral triangle must all have equal lengths, so the perimeter is 4 + 4 + 4 = 12 meters.

Area

There are nice formulas for the areas of some basic shapes. All of the formulas use the same starting point: The area of a rectangle is equal to the product of two adjacent sides. If you label the sides as the **base** (often just the letter b) and **height** (often just the letter h), then this formula is as follows:

$$Area = b \times h$$

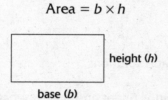

The formula for the area of a triangle requires the length of one side, again called the base, and the length of a line segment extending at right angles from the base to the point opposite, as shown below. The length of this line segment perpendicular to the base is called the height of the triangle. Notice that the height can be measured inside the triangle, but it's also possible to arrange things so that the height lies outside the triangle. What's important is that it must be the length of a line segment perpendicular to the line containing the base of the triangle.

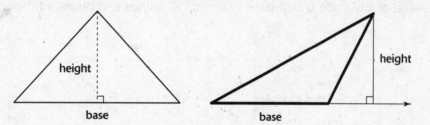

Once you have the length of the base and the height, the area of the triangle is given by the formula:

$$Area = \frac{1}{2} b \times h$$

This makes sense because any triangle can be cut up and rearranged to form a rectangle with the same height and a base half as long.

The area of a parallelogram is given by the formula:

$$Area = b \times h$$

This formula appears exactly like the one for the area of a rectangle, but it is important to realize that, like in a triangle, the height is not necessarily a side of the figure itself, but rather must be measured perpendicular to the base.

The formula for the area of a trapezoid is slightly more complicated than the others mentioned so far. Again we need to measure a height, which is perpendicular to the base, but in this case the length of a single base is not enough. We must have measurements for the two parallel sides, and one of them is called b_1 while the other is called b_2.

$$\text{Area} = \frac{b_1 + b_2}{2} \times h$$

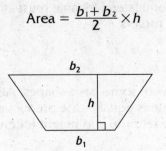

Example Problems

These problems show the answers and solutions.

1. If a rectangle has sides of lengths 5 feet and 3 feet, what is the area of that rectangle?

    answer: 15 square feet

    You multiply the lengths of the base and the height (notice it doesn't matter which of the two adjacent sides you call the base and which the height), so Area = 5 × 3 = 15 square feet.

2. Find the area of a triangle with a base of length 10 inches and height of 7 inches as shown here.

    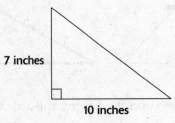

    answer: 35 square inches

    Use the formula, so Area = $\frac{1}{2}$ × 10 × 7 = 35 square inches.

3. Find the area of the triangle shown here.

    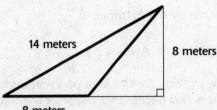

answer: 32 square meters

It is important to recognize that the side of length 14 meters is not needed. The base is 8 meters in length, and the height is also 8 meters. So Area = $\frac{1}{2} \times 8 \times 8 = 32$ square meters.

4. Find the area and perimeter of the parallelogram shown here, where all measurements are given in inches.

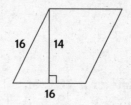

answer: Area = 224 square inches, perimeter = 64 inches

You have a base of length 16 inches, but the other side length of 16 inches is not important for finding the area. The height of 14 inches is what you need, so Area = $16 \times 14 = 224$ square inches.

As for the perimeter, the unmarked sides must match the sides opposite them, so all sides are of length 16 inches, and the perimeter is $16 + 16 + 16 + 16 = 64$ inches.

5. Find the area of the trapezoid shown here, where the measurements shown are in meters.

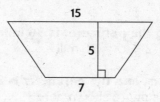

answer: 55 square meters

You see from the picture that $b_1 = 7$ meters, $b_2 = 15$ meters, and $h = 5$ meters. Then Area = $\frac{7 + 15}{2} \times 5 = \frac{22}{2} \times 5 = 11 \times 5 = 55$ square meters.

Work Problems

Use these problems to give yourself additional practice.

1. Find the area and perimeter of the triangle shown, where all the measurements are in feet.

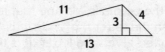

2. Find the area and perimeter of a square with a side of length 10 inches.

3. Find the area and perimeter of a rectangle with sides of lengths 4 centimeters and 6 centimeters.

4. Find the area and perimeter of the parallelogram shown here, where all measurements are in miles.

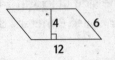

5. Sketch, then find the perimeter and area of, a trapezoid with $b_1 = 10$ centimeters, $b_2 = 16$ centimeters, $h = 4$ centimeters, and the two diagonal sides both of length 5 centimeters.

Worked Solutions

1. **Area is $19\frac{1}{2}$ feet, and the perimeter is 28 feet.** The base has length 13 feet and the height is 3 feet, so Area $= \frac{1}{2} \times 13 \times 3 = \frac{39}{2}$ or $19\frac{1}{2}$ feet. The perimeter is the sum of the three side lengths, or $13 + 11 + 4 = 28$ feet.

2. **Area is 100 square inches, and the perimeter is 40 inches.** A square is a rectangle with all sides equal, so both the base and the height are 10 inches, and Area $= 10 \times 10 = 100$ square inches. The perimeter is the sum of all four sides, or $10 + 10 + 10 + 10 = 40$ inches.

3. **Area is 24 square centimeters, and the perimeter is 20 centimeters.** Area $= 4 \times 6 = 24$ square centimeters. For the perimeter you have to remember that the third and fourth sides will also have lengths 4 and 6, so the perimeter is the sum $4 + 6 + 4 + 6 = 20$ centimeters.

4. **Area is 48 square miles, and the perimeter is 36 miles.** Area $= 12 \times 4 = 48$ square miles. Perimeter $= 12 + 6 + 12 + 6 = 36$ miles.

5. **Area is 52 square centimeters, and the perimeter is 36 centimeters.** Your sketch should look something like the figure shown here.

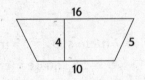

Area $= \frac{10 + 16}{2} \times 4 = \frac{26}{2} \times 4 = 13 \times 4 = 52$ square centimeters, and perimeter $= 10 + 5 + 16 + 5 = 36$ centimeters.

Circles

A **circle** is a figure consisting of all the points in a plane that lie a fixed distance from a particular center point. The **radius** (often given the letter r) of a circle is the distance from the center to any point on the circle. The **diameter** (often given the letter d) is the length of the longest line segment that can be drawn from a point on the circle to another point on the circle, and since this longest line segment must pass through the center, its length will be twice the radius, or $d = 2r$.

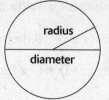

The **circumference** of a circle is the distance around the outside (much like the perimeter of a polygon). The circumference (sometimes given the letter C) is given by either of the following formulas:

$$\text{Circumference} = \pi \times d \text{ or Circumference} = 2\pi \times r$$

The Greek letter pi, which looks like π, is used here to denote a special number. The value of π is approximately 3.14, or $\frac{22}{7}$, but actually pi is an irrational number that cannot be written exactly as a decimal.

The area of a circle is given by the following formula: Area $= \pi \times r^2$.

Example Problems

These problems show the answers and solutions.

1. Find the circumference and area of a circle with a radius of 3 feet.

    answer: 6π feet (circumference) and 9π square feet (area)

    Since you know the radius of the circle, you use the second formula for circumference, and find that Circumference $= 2\pi \times (3) = 6$ feet (or about 18.84 or $18\frac{6}{7}$ feet, approximately). Then using the formula for area, you find that Area $= \pi (3)^2 = 9\pi$ square feet (approximately 28.26 or $28\frac{2}{7}$ square feet).

2. Find the circumference and area of a circle with a diameter of 8 meters.

    answer: 8π meters (circumference) and 16π square meteres (area)

    Notice that this time you were given the diameter rather than the radius. Since $d = 2r$, you have $8 = 2r$, and dividing both sides of this equation by 2 gives you $r = 4$ meters. Then you can use the first formula for circumference to get Circumference $= \pi \times (8) = 8\pi$ meters (approximately 25.12 or $25\frac{1}{7}$ meters), and the formula for area to get Area $= \pi \times (4)^2 = 16\pi$ square meters (approximately 50.24 or $50\frac{2}{7}$ square meters).

Work Problems

1. Find the circumference of a circle with a radius of 1 mile.

2. Find the area of a circle with a radius of 1 mile.

3. Find the circumference of a circle with a diameter of 10 meters.

4. Find the area of a circle with a diameter of 10 meters.

5. Find the area of half of a circle with a radius of 2 feet.

Worked Solutions

1. **2π miles** Circumference $= 2\pi \times (1) = 2\pi$ miles (approximately 6.28 or $6\frac{2}{7}$ miles).

2. **π miles** Area $= \pi\,(1)^2 = \pi$ square miles (approximately 3.14 or $3\frac{1}{7}$ square miles).

3. **10π meters** Circumference $= \pi \times (10) = 10\pi$ meters (approximately 31.4 or $31\frac{3}{7}$ meters).

4. **25π square meters** Since $d = 2r$, or $10 = 2r$, if we divide both sides by 2 we have $r = 5$ meters. Then Area $= \pi \times (5)^2 = 25\pi$ square meters (approximately 78.5 or $78\frac{4}{7}$ square meters).

5. **2π square feet** The entire circle would have Area $= \pi \times (2)^2 = 4\pi$ square feet, so if we wanted the area of half the circle, it would be half that total area, or $\frac{4\pi}{2} = 2\pi$ square feet (approximately 6.28 or $6\frac{2}{7}$ square feet).

Chapter 10
Charts and Graphs

There are many ways of representing information visually. This chapter will introduce some common types: bar charts, circle charts, and line graphs.

Bar Charts

Bar charts are a visual way of presenting several quantities with bars of different lengths. The longer each bar, the greater the quantity being represented. Bar charts are especially effective for representing comparisons over time or between groups of values.

Example Problems

These problems show the answers and solutions.

1. Given the chart shown,

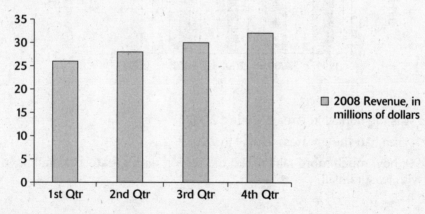

(a) What were the revenues in the first quarter?

(b) If the fourth quarter revenue in 2007 was $22 million, by how much did revenue for the fourth quarter increase from 2007 to 2008?

(c) What were the total 2008 revenues?

(d) What is the overall trend?

answers:

(a) The revenue appears to be about $26 million for the first quarter.

(b) The revenue appears to be about $32 million for the fourth quarter, so this represents an increase of $32 million − $22 million = $10 million from the revenues of the fourth quarter in 2002.

(c) The total revenue is the sum of the revenues for the four quarters, so 26 + 28 + 30 + 32 = $116 million.

(d) The overall trend seems to be a steady increase.

Notice that estimation is often a factor in reading charts and graphs. The actual revenues pictured in the Revenue bar chart are $25.8 million, $28.3 million, $30.1 million, and $32.4 million, but it is extremely difficult to determine these exact amounts from the chart. However, other aspects of the situation, like the steady growth, are more immediately apparent from the chart than they might be from a list of numbers. Any method of presenting information will have advantages and disadvantages, and it is important to be aware of them.

2. Given the chart shown,

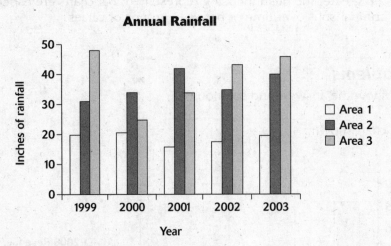

(a) What was the rainfall in Area 3 during 2002?

(b) Which area had the greatest rainfall in 2001?

(c) In 1999, how much more rainfall did the region with greatest rainfall have than the area with least rainfall?

answers:

(a) The rainfall for Area 3 in 2002 appears to be about 43 inches.

(b) Area 2 had the greatest rainfall in 2001.

(c) In 1999, Area 3 appears to have had the most with 48 inches of rainfall, while Area 1 had the least with 20 inches of rainfall, so the difference is 48 − 20 = 28 inches of rainfall.

Work Problems

Use these problems to give yourself additional practice.

Use the chart shown for problems 1 through 3:

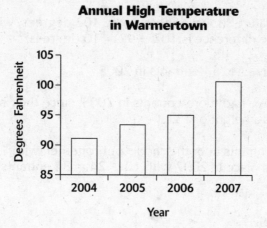

1. What was the high temperature in Warmertown for 2004?

2. For how many of the years shown was the high temperature above 100 degrees?

3. How much higher was the warmest high temperature shown than the lowest high temperature?

Use the chart shown for problems 4 through 6:

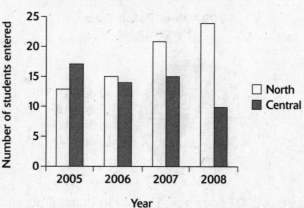

4. How many math contest entrants did Central High School have in 2005?

5. Which high school had more math contest entrants in 2005?

6. In which year were the most students entered in high school math contests at both schools put together?

Worked Solutions

1. The high temperature in Warmertown for 2004 appears to be 92 degrees.

2. For only one of the years (2007) was the high temperature above 100 degrees.

3. The warmest temperature shown appears to be 102 degrees, while the lowest appears to be 92 degrees, so the difference is $102 - 92 = 10$ degrees.

4. Central appears to have had 14 entrants in 2005.

5. Central appears to have had more entrants in 2005, since their bar has a height of 17 while North's bar has a height of 13.

6. 2006 had the most entrants at both schools put together, with $15 + 21 = 36$ entrants (compared to second place in 2007 with $10 + 24 = 34$ entrants).

Circle Charts

Circle charts (often called **pie charts**) are a visual way of showing how a total amount is broken down into parts. The larger the sector of the circle, the larger the share of the whole that component represents. There are many different variations in the style and labeling of circle charts, so it is important to pay close attention to labels and all other information given to avoid misunderstandings.

Example Problem

1. Based on the chart shown,

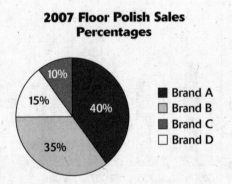

2007 Floor Polish Sales Percentages

- Brand A
- Brand B
- Brand C
- Brand D

(a) What percentage of 2007 floor polish sales did Brand C account for?

(b) How much larger was the percentage of the leading brand than the second-largest brand?

(c) What percentage of 2007 floor polish sales did the three largest brands account for?

(d) If total floor polish sales were 65,000 units for 2007, how many units did Brand D account for?

answers:

(a) According to the chart, Brand C accounted for 15% of 2007 floor polish sales.

(b) The leading brand had 40%, whereas the second-largest brand had 35%, so the leading brand's sales were larger by 5%.

(c) The three largest brands accounted for 40%, 35%, and 15%, for a total of 40 + 35 + 15 = 90% of the sales.

(d) Brand D had 10% of the total sales, and 10% of 65,000 is 0.10 (65000 = 6500 units).

Work Problems

Use these problems to give yourself additional practice.

Use the chart shown for problems 1–4:

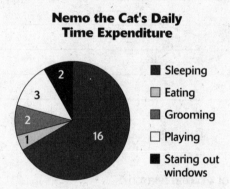

**Nemo the Cat's Daily
Time Expenditure**

- Sleeping
- Eating
- Grooming
- Playing
- Staring out windows

1. How many hours a day does Nemo spend sleeping?

2. How many hours a day does Nemo spend playing and staring out windows combined?

3. How many hours does Nemo sleep for each hour he eats?

4. When Nemo was a kitten, he slept only about three quarters as much as he does now, and spent the extra time playing. How much time a day did Nemo spend playing when he was a kitten?

Worked Solutions

1. Nemo spends 16 hours a day sleeping.

2. Nemo spends 3 hours playing and 2 hours staring out windows, so 3 + 2 = 5 hours combined.

3. Nemo spends 16 hours sleeping for every 1 hour eating, or $\frac{16}{1}$ = 16 hours sleeping for each hour eating.

4. Now Nemo sleeps 16 hours each day, so as a kitten he slept $\frac{3}{4} \times 16 = 12$ hours a day. Since this is 16 − 12 = 4 hours less than he sleeps each day now, adding this to his play time means as a kitten he spent 3 + 4 = 7 hours a day playing.

Line Graphs

Line graphs are a visual way of presenting a series of quantities. Each piece of data is plotted as a point, and usually the points are then connected. Line graphs can be especially useful for showing connections or trends, but in many ways they are like bar charts.

Example Problems

This problem shows the answer and solution.

1. Based on the graph shown,

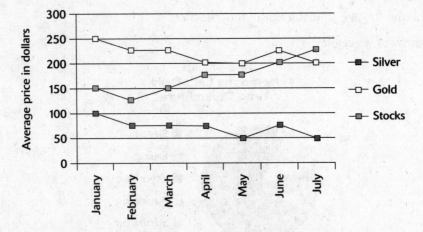

   (a) What was the price of gold in March?

   (b) How much did the price of stocks change from April to July?

   (c) If you spent $1000 on silver in January and sold all of it in July, how much money would you receive from the sale?

   (d) Overall, which of gold, silver, and stocks perform most alike?

answers:

(a) The price of gold in March appears to be $225.

(b) The price of stocks was $175 in April and $225 in July, so the change was an increase of $225 - 175 = 50.

(c) In January the price of silver was $100, so spending $1000 could buy $\frac{1000}{100} = 10$ units of silver. In July the price of silver was $50, so selling your 10 units would yield $50 \times 10 = 500.

(d) Overall, it appears that gold and silver prices move in very similar ways, while stock prices move sometimes with but sometimes opposite the prices of gold and silver.

Work Problems

Use the graph shown for problems 1−5:

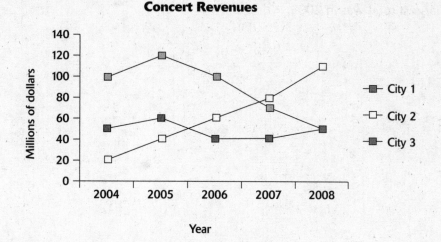

Concert Revenues

1. What were the concert revenues in City 2 for 2005?

2. Which city had the largest concert revenues in 2007?

3. Between which two years did concert revenues in City 1 change the most?

4. What was the difference between the largest and smallest revenues of City 3?

5. In which year were the combined concert revenues of the three cities largest?

Worked Solutions

1. The concert revenues in City 2 for 2005 were $40 million.

2. City 2 had the largest concert revenues in 2007.

3. In 2007, City 1's concert revenues were $70 million, compared to $100 million in 2006. This represents a decrease of $100 − 70 = $30 million. So, the revenues changed the most between 2006 and 2007.

4. Since City 3's largest revenues were $60 million and its smallest revenues were $40 million, the difference was $60 − 40 = $20 million.

5. We add the revenues for the three cities for each year:

   For 2004, $20 + 50 + 100 = $170 million

   For 2005, $40 + 60 + 120 = $220 million

   For 2006, $40 + 60 + 100 = $200 million

For 2007, 40 + 70 + 80 = $190 million

For 2008, 50 + 50 + 110 = $210 million

So the largest total was in 2005.

Chapter 11
Probability and Statistics

Probability is the study of the chances of certain events occurring, and statistics can be used to describe characteristics of large bodies of information. We will introduce elements of both in this chapter.

Probability

Probability is a mathematical measurement of the likelihood of a certain event occurring. Technically, **probability** can be defined as

$$\text{probability} = \frac{\text{number of favorable events}}{\text{number of total events}}$$

This definition is best understood through some examples.

Example Problems

These problems show the answers and solutions.

1. When a fair coin is flipped, what is the probability of the coin coming up heads?

    answer: $\frac{1}{2}$

    There is one event (the coin coming up heads) which is considered favorable, out of a total of two events (the coin coming up heads and the coin coming up tails), and thus a probability of $\frac{1}{2}$ of a result of heads.

    Notice that when we say "a fair coin" we automatically mean one for which either face is equally likely to come up, which isn't necessarily always the case in the real world.

2. If a jar contains three balls, one red, one blue, and one green, and a ball is picked at random from the jar, what is the probability that the ball picked is red?

    answer: $\frac{1}{3}$

    There is one event (the red ball being picked), that is considered favorable, out of a total of three events (the red ball being picked, the blue ball being picked, and the green ball being picked), so the probability of the red ball being picked is $\frac{1}{3}$.

3. If a standard die is tossed, what is the probability of rolling 2 or less?

    answer: $\frac{2}{6}$ or $\frac{1}{3}$

    Since a standard die has six faces, there are six events possible (rolling a 1, rolling a 2, rolling a 3, rolling a 4, rolling a 5, and rolling a 6). Out of these, there are two favorable events (rolling a 1 and rolling a 2). Then, the probability of rolling a 2 or less is $\frac{2}{6}$ or $\frac{1}{3}$.

4. If a card is drawn at random out of a standard deck (not including jokers), what is the probability of drawing a king?

    answer: $\frac{4}{52}$ or $\frac{1}{13}$

    There are 52 cards in a standard deck that could be drawn (an ace, 2, 3, 4, 5, 6, 7, 8, 9, 10, jack, queen, and king in each of four suits). Of these, there are four that count as a favorable event (the kings of clubs, diamonds, hearts, and spades), so the probability of drawing a king is $\frac{4}{52}$ or $\frac{1}{13}$.

Work Problems

Use these problems to give yourself additional practice.

1. If a jar contains four balls, one red, one blue, one white, and one green, and a ball is picked at random, what is the probability that the ball picked is green?

2. If a standard die is tossed, what is the probability of rolling a 2 or more?

3. A child has seven coins in a piggy bank: a quarter, two dimes, a nickel, and three pennies. If one coin is taken randomly out of the bank, what is the probability that coin is the quarter?

4. If a card is drawn at random out of a standard deck (not including jokers), what is the probability of drawing a numbered card 2 through 10?

5. A jar contains 500 jelly beans, 100 of them red, 120 of them black, 60 of them green, 130 of them orange, and 90 of them purple. Jon loves red and orange jelly beans, but hates black ones. If Jon randomly draws a jelly bean from the jar, what is the probability of getting one he loves?

Worked Solutions

1. $\frac{1}{4}$ There is one favorable event (drawing the green ball) out of four possible events total (drawing red, drawing blue, drawing white, and drawing green), so the probability of drawing the green ball is $\frac{1}{4}$.

2. $\frac{5}{6}$ There are five favorable events (rolling a 2, rolling a 3, rolling a 4, rolling a 5, or rolling a 6) out of a total of six events (rolling a 1, rolling a 2, rolling a 3, rolling a 4, rolling a 5, or rolling a 6), so the probability of rolling a 2 or more is $\frac{5}{6}$.

3. $\frac{1}{7}$ There is one favorable event (drawing the quarter) out of seven total events (one for each of the seven coins in the bank—notice that the dimes count twice and the pennies count three times, since there are several of them that could be drawn), so the probability of drawing the quarter is $\frac{1}{7}$.

4. $\frac{36}{52} = \frac{9}{13}$ There are 36 favorable events (drawing the 2, 3, 4, 5, 6, 7, 8, 9, or 10 in each of four suits) out of 52 total events, so the probability of drawing a numbered card is $\frac{36}{52} = \frac{9}{13}$.

5. $\frac{230}{500} = \frac{23}{50}$ There are 230 favorable events (100 possibilities of drawing red jelly beans plus 130 possibilities of drawing orange ones) out of a total of 500 events (one for each jelly bean that could be drawn), for a probability of $\frac{230}{500} = \frac{23}{50}$.

Independent Events

Probabilities of some events depend on previous events, for instance drawing a king from a deck of cards after three other kings have been drawn without replacing them. On the other hand, some events have equal likelihoods regardless of previous events, for instance the probability of a fair coin coming up heads immediately after it has come up tails. When two events have no influence on each other's chances, they are called **independent events**. The probability of two independent events both occurring is the product of the probability of the first happening times the probability of the second happening.

Example Problems

These problems show the answers and solutions.

1. A jar contains three balls, one red, one blue, and one green, and two balls are picked at random, with the ball drawn being replaced after each draw. What is the probability that the first ball picked is red but the second ball picked is not red?

   answer: $\frac{2}{9}$

   You have the probability of the first ball picked being red as $\frac{1}{3}$ (see Example Problem 2 in the section, "Probability"). Then when the first ball drawn is replaced and another ball is drawn, the probability of it not being red is $\frac{2}{3}$ (two favorable events, blue and green, out of three total events possible). Then the probability of both these events occurring is equal to their product, or $\frac{1}{3} \times \frac{2}{3} = \frac{2}{9}$.

2. A fair coin is flipped twice. What is the probability of heads coming up both times?

   answer: $\frac{1}{4}$

   The probability of heads coming up on the first toss is $\frac{1}{2}$, and the probability of heads coming up on the second toss is also $\frac{1}{2}$, so the probability of both these events occurring is $\frac{1}{2} \times \frac{1}{2} = \frac{1}{4}$.

3. A standard die is tossed twice. What is the probability that the first roll will be either a 3 or 4, but the second roll will be less than 3?

   answer: $\frac{1}{9}$

The probability of the first roll being a 3 or 4 is $\frac{2}{6} = \frac{1}{3}$, and the probability of the second roll being a 1 or 2 is also $\frac{2}{6} = \frac{1}{3}$. Then the probability of both of these events occurring is $\frac{1}{3} \times \frac{1}{3} = \frac{1}{9}$.

Work Problems

Use these problems to give yourself additional practice.

1. A jar contains three white marbles and two black marbles. A marble is drawn, then replaced, and then a second marble is drawn. What is the probability that both marbles are black?

2. With the jar from Problem 1, what is the probability that both marbles drawn are white?

3. A card is drawn randomly from a standard deck, then replaced, and a second card is then drawn. What is the probability that the cards are both spades?

4. A fair coin is flipped twice. What is the probability of heads on the first toss, but tails on the second?

5. A standard die is tossed three times. What is the probability of rolling three sixes?

Worked Solutions

1. $\frac{4}{25}$ The probability that the first marble drawn is black is $\frac{2}{5}$, and the probability that the second marble drawn is black is also $\frac{2}{5}$, so the probability of both events happening is $\frac{2}{5} \times \frac{2}{5} = \frac{4}{25}$.

2. $\frac{9}{25}$ The probability that the first marble drawn is white is $\frac{3}{5}$, and the probability that the second marble drawn is white is also $\frac{3}{5}$, so the probability of both events happening is $\frac{3}{5} \times \frac{3}{5} = \frac{9}{25}$.

3. $\frac{1}{16}$ Since 13 of the 52 cards in a standard deck are spades, the probability of drawing one is $\frac{13}{52}$ or $\frac{1}{4}$ for the first draw, and $\frac{1}{4}$ for the second draw. Then the probability of both cards drawn being spades is $\frac{1}{4} \times \frac{1}{4} = \frac{1}{16}$.

4. $\frac{1}{4}$ The probability of heads on the first toss is $\frac{1}{2}$, and the probability of tails on the second toss is also $\frac{1}{2}$, so the probability of both events occurring is $\frac{1}{2} \times \frac{1}{2} = \frac{1}{4}$.

5. $\frac{1}{216}$ The probability of a 6 on each individual roll is $\frac{1}{6}$, so the probability of all three rolls being sixes is $\frac{1}{6} \times \frac{1}{6} \times \frac{1}{6} = \frac{1}{216}$.

More Complicated Probabilities

There are many other rules for probabilities that will not be covered here, but many complicated situations can be addressed by dividing the number of favorable events by the total number of possible events, as long as we can find ways to count these.

Example Problems

These problems show the answers and solutions.

1. What is the probability of getting exactly two heads on three tosses of a fair coin?

    You can list out the possible results like this:

    If heads is thrown first: $\begin{cases} \text{HHH} & \boxed{\text{HHT}} \\ \boxed{\text{HTH}} & \text{HTT} \end{cases}$

    If tails is thrown first: $\begin{cases} \boxed{\text{THH}} & \text{THT} \\ \text{TTH} & \text{TTT} \end{cases}$

    answer: $\frac{3}{8}$

    Since there are three possibilities (the ones indicated with boxes) listed that include exactly two heads, from among eight possible outcomes, the probability is $\frac{3}{8}$.

2. If two standard dice are rolled and the results are added, what is the probability of a resulting total less than 5?

    You can list out all possible rolls for both dice as follows:

    $$\begin{array}{cccccc}
    \boxed{1,1} & \boxed{1,2} & \boxed{1,3} & 1,4 & 1,5 & 1,6 \\
    \boxed{2,1} & \boxed{2,2} & 2,3 & 2,4 & 2,5 & 2,6 \\
    \boxed{3,1} & 3,2 & 3,3 & 3,4 & 3,5 & 3,6 \\
    4,1 & 4,2 & 4,3 & 4,4 & 4,5 & 4,6 \\
    5,1 & 5,2 & 5,3 & 5,4 & 5,5 & 5,6 \\
    6,1 & 6,2 & 6,3 & 6,4 & 6,5 & 6,6
    \end{array}$$

    answer: $\frac{1}{6}$

    Since there are six pairs that total less than 5 (the 1,1; 1,2; 1,3; 2,1; 2,2; and 3,1 pairs marked with boxes) out of a total of 36 possible pairs of rolls, the probability of rolling a total less than a 5 is $\frac{6}{36} = \frac{1}{6}$.

3. A jar contains a red ball, a yellow ball, and a purple ball. If a ball is drawn and replaced, then a second ball is drawn, what is the probability that the balls drawn are different colors?

    You can list out all possible pairs of draws as follows:

    $$\begin{array}{ccc}
    \text{RR} & \boxed{\text{RY}} & \boxed{\text{RP}} \\
    \boxed{\text{YR}} & \text{YY} & \boxed{\text{YP}} \\
    \boxed{\text{PR}} & \boxed{\text{PY}} & \text{PP}
    \end{array}$$

    answer: $\frac{2}{3}$

You see that six of these pairs involve different colors (the ones marked with boxes), out of a total of nine pairs, so the probability is $\frac{6}{9} = \frac{2}{3}$.

Notice that in all of these cases we have been systematic about the order in which we have listed the possible events. It is important to be careful to list all of the possible events and not accidentally overlook one or include duplicates, but no single universal way exists to accomplish this. Being organized about how you list the possible events (like by first listing all events where the first result is heads, for instance) is the best idea, but adapting that to individual situations is up to you.

Work Problems

Use these problems to give yourself additional practice.

1. If two standard dice are tossed, what is the probability that the sum of the two rolls is 9 or more?

2. What is the probability of getting at least one tail on three tosses of a fair coin?

3. If two standard dice are tossed, what is the probability that the first roll is greater than the second?

4. A jar contains a red ball, a yellow ball, a purple ball, and a black ball. If a ball is drawn and replaced, then a second ball is drawn, what is the probability that the balls drawn are different colors?

5. A piggy bank contains a quarter, two dimes, a nickel, and three pennies. If one coin is taken randomly out of the bank, then replaced, and then a second coin is drawn, what is the probability that the total value drawn is at least 30 cents?

Worked Solutions

1. $\frac{5}{18}$ List out all possible pairs of rolls:

$$
\begin{array}{cccccc}
1,1 & 1,2 & 1,3 & 1,4 & 1,5 & 1,6 \\
2,1 & 2,2 & 2,3 & 2,4 & 2,5 & 2,6 \\
3,1 & 3,2 & 3,3 & 3,4 & 3,5 & \boxed{3,6} \\
4,1 & 4,2 & 4,3 & 4,4 & \boxed{4,5} & \boxed{4,6} \\
5,1 & 5,2 & 5,3 & \boxed{5,4} & \boxed{5,5} & \boxed{5,6} \\
6,1 & 6,2 & \boxed{6,3} & \boxed{6,4} & \boxed{6,5} & \boxed{6,6} \\
\end{array}
$$

You count 10 of the 36 possible results that involve totals of 9 or more (the ones marked with boxes), so the probability is $\frac{10}{36} = \frac{5}{18}$.

2. $\frac{7}{8}$ List all possible results:

If heads is thrown first: $\begin{cases} \text{HHH} & \boxed{\text{HHT}} \\ \boxed{\text{HTH}} & \boxed{\text{HTT}} \end{cases}$

If tails is thrown first: $\begin{cases} \boxed{\text{THH}} & \boxed{\text{THT}} \\ \boxed{\text{TTH}} & \boxed{\text{TTT}} \end{cases}$

There are seven that include at least one tail (boxed previously), out of eight possible outcomes, so the probability is $\frac{7}{8}$.

3. $\frac{5}{12}$ Examine the list of possible outcomes:

$$
\begin{array}{cccccc}
1,1 & 1,2 & 1,3 & 1,4 & 1,5 & 1,6 \\
\boxed{2,1} & 2,2 & 2,3 & 2,4 & 2,5 & 2,6 \\
\boxed{3,1} & \boxed{3,2} & 3,3 & 3,4 & 3,5 & 3,6 \\
\boxed{4,1} & \boxed{4,2} & \boxed{4,3} & 4,4 & 4,5 & 4,6 \\
\boxed{5,1} & \boxed{5,2} & \boxed{5,3} & \boxed{5,4} & 5,5 & 5,6 \\
\boxed{6,1} & \boxed{6,2} & \boxed{6,3} & \boxed{6,4} & \boxed{6,5} & 6,6
\end{array}
$$

You see that in 15 of these 36 possible events (the boxed ones), the first roll is greater than the second, so the probability is $\frac{15}{36} = \frac{5}{12}$.

4. $\frac{3}{4}$ List out all possible pairs of draws as follows:

$$
\begin{array}{cccc}
\text{RR} & \boxed{\text{RY}} & \boxed{\text{RP}} & \boxed{\text{RB}} \\
\boxed{\text{YR}} & \text{YY} & \boxed{\text{YP}} & \boxed{\text{YB}} \\
\boxed{\text{PR}} & \boxed{\text{PY}} & \text{PP} & \boxed{\text{PB}} \\
\boxed{\text{BR}} & \boxed{\text{BY}} & \boxed{\text{BP}} & \text{BB}
\end{array}
$$

Then 12 out of the 16 possible results involve balls of different colors, so the probability is $\frac{12}{16} = \frac{3}{4}$.

5. $\frac{1}{7}$ List out all possible pairs of draws:

$$
\begin{array}{ccccccc}
\boxed{\text{QQ}} & \boxed{\text{QD}} & \boxed{\text{QD}} & \boxed{\text{QN}} & \text{QP} & \text{QP} & \text{QP} \\
\boxed{\text{DQ}} & \text{DD} & \text{DD} & \text{DN} & \text{DP} & \text{DP} & \text{DP} \\
\boxed{\text{DQ}} & \text{DD} & \text{DD} & \text{DN} & \text{DP} & \text{DP} & \text{DP} \\
\boxed{\text{NQ}} & \text{ND} & \text{ND} & \text{NN} & \text{NP} & \text{NP} & \text{NP} \\
\text{PQ} & \text{PD} & \text{PD} & \text{PN} & \text{PP} & \text{PP} & \text{PP} \\
\text{PQ} & \text{PD} & \text{PD} & \text{PN} & \text{PP} & \text{PP} & \text{PP} \\
\text{PQ} & \text{PD} & \text{PD} & \text{PN} & \text{PP} & \text{PP} & \text{PP}
\end{array}
$$

Seven of these pairs (the ones boxed above, QQ, QD, QD, QN, DQ, DQ, and NQ) total at least 30 cents, out of a total of 49 pairs, so the probability is $\frac{7}{49} = \frac{1}{7}$.

Statistics

The science of statistics studies numerical data. One thing statistics is useful for is summarizing large amounts of data. Some of the most common ways of summarizing include the mean, median, and mode.

The **mean** (often called the **arithmetic mean** or **average**) of a collection of values is the sum of the values, divided by how many there are.

If a collection of values is listed in order, the **median** is the middle value (if the number of values is odd), or the average of the two middle values (if the number of values is even). Both means and medians indicate something about what a "typical" value in the collection is like, but medians are not as influenced by the largest and smallest values in the collection.

The **mode** of a collection of values is the value that occurs the most times.

Example Problems

These problems show the answers and solutions.

1. What is the mean of 7, 9, and 14?

   answer: 10

   Add the three values, $7 + 9 + 14 = 30$, and then divide by 3 because that's how many values there are, so $\frac{30}{3} = 10$ is the mean.

2. What is the mean of 24, 50, 16, 20, and 30?

   answer: 28

   Add the five values, $24 + 50 + 16 + 20 + 30 = 140$, and then divide by 5, so $\frac{140}{5} = 28$ is the mean.

3. What is the median of 24, 50, 16, 20, and 30?

   answer: 24

   Arrange the values in order: 16, 20, 24, 30, 50. Now the middle value is 24, so 24 is the median.

4. What is the median of 4, 11, 43, 23, 11, and 30?

   answer: 17

   Arrange the values in order: 4, 11, 11, 23, 30, 43. Now 11 and 23 are the middle two values, so the median is their mean: $\frac{11 + 23}{2} = 17$.

5. What is the mode of 4, 11, 43, 23, 11, and 30?

   answer: 11

   Since 11 occurs twice, more than any of the other values, it is the mode.

6. What is the mode of 6, 4, 5, 9, 3, 4, 8, 3, 9, and 4?

answer: 4

Since 4 occurs three times, more than any of the other values, it is the mode.

Work Problems
Use these problems to give yourself additional practice.

1. What are the mean, median, and mode of 3, 3, 6, and 8?

2. What are the mean, median, and mode of 13, 15, 18, 18, and 14?

3. What are the mean, median, and mode of 50, 60, 20, 40, 50, 20, 60, 10, 60, 70, and 90?

Worked Solutions

1. **Mean is 5; median is 4.5; mode is 3.** The mean is the sum of the four values divided by how many values there are, or $\frac{3+3+6+8}{4} = \frac{20}{4} = 5$.

Since the values are already in order, the median is the mean of the two middle values, or (3 + 6)/2 = 4.5.

The mode is 3 since it is the value that occurs most often.

2. **Mean is 15.6; median is 15; mode is 18.** The mean is $\frac{13+15+18+18+14}{5} = \frac{78}{5} = 15.6$.

To find the median we must first arrange the values in order: 13, 14, 15, 18, 18. Then the median is 15 since it is the middle value.

The mode is 18 since it is the value that occurs most often.

3. **Mean is 48.$\overline{18}$; median is 50; mode is 60.** The mean is $\frac{50+60+20+40+50+20+60+10+60+70+90}{11} = 48.\overline{18}$.

To find the median we first arrange the values in order: 10, 20, 20, 40, 50, 50, 60, 60, 60, 70, 90. Then the median is 50 since it is the middle value.

The mode is 60 since it is the value that occurs most often.

Chapter 12
Variables, Algebraic Expressions, and Simple Equations

Variables and Algebraic Expressions

A general knowledge of variables and expressions is needed before you begin solving equations.

Variables

A **variable** is a letter that is used to represent an unknown quantity. The most commonly used variables are a, b, c, m, n, x, y, and z. Usually if nothing else is specified, you use the letter x as your variable. Certain letters, such as e, i, l, and o, are seldom used as variables because they have special meanings in algebra or can be mistaken for one or zero.

Algebraic Expressions

You can use variables to translate verbal expressions into **algebraic expressions,** where you use letters to stand for numbers. Consider these key terms when translating words into letters and numbers:

❑ **For addition:** sum, greater than, increased by, more than

❑ **For subtraction:** difference, minus, smaller than, less than, decreased by

❑ **For multiplication:** product, times, multiplied by

❑ **For division:** quotient, halve, divided by, ratio

Example Problems

These problems show the answers and solutions.

1. Give an algebraic expression for the sum of three and a number.

   answer: One possible answer is $3 + x$.

2. Give an algebraic expression for six more than a number.

   answer: One possible answer is $a + 6$.

3. Give an algebraic expression for a number minus eight.

   answer: One possible answer is $y - 8$.

4. Give an algebraic expression for ten less than a number.

   answer: One possible answer is $b - 10$.

5. Give an algebraic expression for twice a number.

   answer: Twice means two times, so one possible answer is $2m$.

6. Give an algebraic expression for a number multiplied by five.

   answer: One possible answer is $5c$.

7. Give an algebraic expression for a number divided by seven.

   answer: One possible answer is $\frac{x}{7}$.

8. Give an algebraic expression for half of a number.

   answer: One possible answer is $\frac{n}{2}$.

Work Problems

Use these problems to give yourself additional practice.

1. Give an algebraic expression for one more than a number.

2. Give an algebraic expression for five less than a number.

3. Give an algebraic expression for three times a number.

4. Give an algebraic expression for a third of a number.

Worked Solutions

1. $1 + y$

2. $c - 5$

3. $3b$

4. $\dfrac{m}{3}$

Evaluating Expressions

To **evaluate an expression**, replace the variable(s) with the value (or values) given for the specific variable (or variables) and then do the arithmetic. Remember to simplify the expression using the correct order of operations: parentheses, exponents, multiplication/division, and addition/subtraction.

Example Problems

These problems show the answers and solutions.

1. Evaluate $2 + 3x$ for $x = 4$.

   answer: 14

$$2 + 3x = 2 + 3(4)$$
$$= 2 + 12$$
$$= 14$$

2. Evaluate $4x + y$ for $x = 2$ and $y = 1$.

   answer: 9

$$4x + y = 4(2) + (1)$$
$$= 8 + 1$$
$$= 9$$

3. Evaluate $\dfrac{m}{5} - n$ for $m = 10$ and $n = 1$.

   answer: 1

$$\frac{m}{5} - n = (10)/5 - (1)$$
$$= 2 - 1$$
$$= 1$$

4. Evaluate $6m + \dfrac{8}{n}$ for $m = 2$ and $n = 4$.

answer: 14

$$6m + \frac{8}{n} = 6(2) + \frac{8}{(4)}$$
$$= 12 + 2$$
$$= 14$$

5. Evaluate $\dfrac{(a + 2b)}{3}$ for $a = 6$ and $b = 3$.

answer: 4

$$\frac{(a + 2b)}{3} = \frac{[(6) + 2(3)]}{3}$$
$$= \frac{(6 + 6)}{3}$$
$$= \frac{12}{3}$$
$$= 4$$

6. Evaluate $x^2 + 5y$ for $x = 3$ and $y = 2$.

answer: 19

$$x^2 + 5y = (3)^2 + 5(2)$$
$$= 9 + 10$$
$$= 19$$

7. Evaluate $b\left(4 - \dfrac{c}{3}\right)$ for $b = 6$ and $c = 12$.

answer: 0

$$b\left(4 - \frac{c}{3}\right) = (6)\left(4 - \frac{12}{3}\right)$$
$$= 6(4 - 4)$$
$$= 6(0)$$
$$= 0$$

8. Evaluate $(5 + x)\left(\dfrac{y^3}{4} + 1\right)$ for $x = 3$ and $y = 2$.

answer: 24

$$(5 + x)\left(\frac{y^3}{4} + 1\right) = (5 + 3)\left(\frac{2^3}{4} + 1\right)$$
$$= (8)\left(\frac{8}{4} + 1\right)$$
$$= (8)(2 + 1)$$
$$= (8)(3)$$
$$= 24$$

Work Problems

Use these problems to give yourself additional practice.

1. Evaluate $2 + 3x$ for $x = 5$.

2. Evaluate $\frac{a}{5} - b$ for $a = 15$ and $b = 3$.

3. Evaluate $\frac{(a + 4b)}{3}$ for $a = 2$ and $b = 4$.

4. Evaluate $5y^2 - 1$ for $y = 3$.

5. Evaluate $m\left(4 - \frac{n}{6}\right)$ for $m = 2$ and $n = 12$.

Worked Solutions

1. **17**

$$2 + 3x = 2 + 3(5)$$
$$= 2 + 15$$
$$= 17$$

2. **0**

$$\frac{a}{5} - b = \frac{(15)}{5} - (3)$$
$$= 3 - 3$$
$$= 0$$

3. **6**

$$\frac{(a + 4b)}{3} = \frac{[(2) + 4(4)]}{3}$$
$$\frac{[2 + 16]}{3} = \frac{18}{3}$$
$$= 6$$

4. **44**

$$5y^2 - 1 = 5(3)^2 - 1$$
$$= 5(9) - 1$$
$$= 45 - 1$$
$$= 44$$

5. **4**

$$m\left(4 - \frac{n}{6}\right) = 2\left(4 - \frac{(12)}{6}\right)$$
$$= 2(4 - 2)$$
$$= 2(2)$$
$$= 4$$

Solving Simple Equations

When **solving a simple equation**, it is important to think of the equation as a balance with the center being the equal sign (=). This means that what you do to one side of the equation must also be done to the other side. This keeps the equation balanced. An example of this would be adding 6 to each side. Solving an equation involves isolating the variable you want to solve for. What you're looking for should be on one side of the equal sign, while everything else is on the other side. If you're solving for the variable y, you want to get y by itself on one side and everything else on the other side of the equal sign.

A **solution** to an equation is a value for the variable that makes the equation true.

Example Problems

These problems show the answers and solutions.

1. Is $x = 4$ a solution to the equation $x + 5 = 9$?

    answer: Yes.   If you put 4 in the x spot of this equation, you get the following:

    $$(4) + 5 = 9$$
    $$9 = 9 \ \checkmark$$

    This is a true statement, so $x = 4$ is a solution to this equation. The check mark indicates that the solution has been verified.

2. Is $y = 3$ a solution to the equation $\frac{y}{6} = 2$?

    answer: No.   If you substitute 3 for y in this equation, you get the following:

    $$\frac{3}{6} = 2$$
    $$\frac{1}{2} = 2$$

    This is not a true statement, so $y = 3$ is not a solution to this equation.

Solving Addition and Subtraction Equations

These are equations that involve only addition or subtraction or a combination of both.

Example Problems

These problems show the answers and solutions.

1. Solve $3 + y = 8$ for y.

    answer: $y = 5$

    This equation is solved by getting y by itself. This is done by using opposite operations. Because the equation involves addition, you solve it by subtracting. Thus, you subtract 3 from both sides:

$$3 + y = 8$$
$$\underline{-3 \quad -3}$$
$$y = 5$$

To check your answer, put the answer back into the equation:

$$3 + y = 8$$
$$3 + 5 = 8$$
$$8 = 8 \quad \checkmark$$

Because $8 = 8$ is true, your solution $y = 5$ is correct.

2. Solve $x - 4 = 6$ for x.

    answer: $x = 10$

$$x - 4 = \quad 6$$
$$\underline{+4 \quad + \ 4}$$
$$x = 10$$

To check your answer, substitute 10 for x in the original equation:

$$(10) - 4 = 6$$
$$6 = 6 \quad \checkmark$$

3. Solve $12 + a = 20$ for a.

    answer: $a = 8$

$$12 + a = 20$$
$$\underline{-12 \quad -12}$$
$$a = 8$$

To check:

$$12 + (8) = 20$$
$$20 = 20 \quad \checkmark$$

4. Solve $z - 30 = 5$ for z.

    answer: $z = 35$

$$z - 30 = 5$$
$$\underline{+30 \quad +30}$$
$$z = 35$$

To check:

$$(35) - 30 = 5$$
$$5 = 5 \quad \checkmark$$

Multiplication and Division Equations

These are equations that involve only multiplication or division. The variable is usually already on one side of the equation, but there it is either a fraction of the variable or more than one of the variables (like $\frac{x}{2}$ or $5x$, for instance). Multiplication and division of equations are similar to addition and subtraction; what is done to one side of the equation must be done to the other side, as long as you are not multiplying or dividing by zero.

Example Problems

These problems show the answers and solutions.

1. Solve $8a = 16$ for a.

    answer: $a = 2$

    To solve, divide each side by 8:

    $$8a = 16$$
    $$\frac{8a}{8} = \frac{16}{8}$$
    $$\frac{\cancel{8}a}{\cancel{8}} = \frac{16}{8}$$
    $$a = 2$$

    To check, substitute $a = 2$ back into the equation:

    $$8a = 16$$
    $$8(2) = 16$$
    $$16 = 16 \quad \checkmark$$

2. Solve $5y = 10$ for y.

    answer: $y = 2$

    $$5y = 10$$
    $$\frac{5y}{5} = \frac{10}{5}$$
    $$\frac{\cancel{5}y}{\cancel{5}} = \frac{10}{5}$$
    $$y = 2$$

To check:

$$5(2) = 10$$
$$10 = 10 \quad \checkmark$$

3. Solve $6x = 3$ for x.

answer: $x = \dfrac{1}{2}$

$$6x = 3$$
$$\frac{6x}{6} = \frac{3}{6}$$
$$\frac{\cancel{6}x}{\cancel{6}} = \frac{3}{6}$$
$$x = \frac{1}{2}$$

To check:

$$6(1/2) = 3$$
$$3 = 3 \quad \checkmark$$

4. Solve $\dfrac{b}{2} = 4$ for b.

answer: $b = 8$

This time the variable is divided by 2, so in order to offset this you have to multiply both sides of the equation by 2:

$$\frac{b}{2} = 4$$
$$2 \cdot \frac{b}{2} = 2 \cdot 4$$
$$\cancel{2} \cdot \frac{b}{\cancel{2}} = 8$$
$$b = 8$$

To check:

$$(8)/2 = 4$$
$$4 = 4 \quad \checkmark$$

5.   Solve $\frac{x}{5} = 6$ for x.

    answer: $x = 30$

$$\frac{x}{5} = 6$$

$$5 \cdot \frac{x}{5} = 5 \cdot 6$$

$$\cancel{5} \cdot \frac{x}{\cancel{5}} = 30$$

$$x = 30$$

    To check:

$$\frac{(30)}{5} = 6$$

$$6 = 6 \quad \checkmark$$

6.   Solve $\frac{2y}{3} = 8$ for y.

    answer: $y = 12$

This time the variable is being multiplied by $\frac{2}{3}$, so to undo this you need to multiply both sides of the equation by $\frac{3}{2}$:

$$\frac{2y}{3} = 8$$

$$\frac{3}{2} \cdot \frac{2y}{3} = \frac{3}{2} \cdot 8$$

$$\frac{\cancel{3}}{\cancel{2}} \cdot \frac{\cancel{2}y}{\cancel{3}} = \frac{3}{\cancel{2}} \cdot \overset{4}{\cancel{8}}$$

$$y = 12$$

    To check:

$$\frac{2(12)}{3} = 8$$

$$\frac{24}{3} = 8$$

$$8 = 8 \quad \checkmark$$

7.   Solve $\frac{a}{4} = \frac{5}{2}$ for a.

    answer: $a = 10$

$$\frac{a}{4} = \frac{5}{2}$$

$$4 \cdot \frac{a}{4} = 4 \cdot \frac{5}{2}$$

$$\cancel{4} \cdot \frac{a}{\cancel{4}} = \overset{2}{\cancel{4}} \cdot \frac{5}{\cancel{2}}$$

$$a = 10$$

To check:

$$\frac{10}{4} = \frac{5}{2}$$

$$\frac{5}{2} = \frac{5}{2} \quad \checkmark$$

Work Problems

Use these problems to give yourself additional practice.

1. Is $a = 5$ a solution to the equation $4 + a = 5$?

2. Solve the equation $3 + x = 9$ for x.

3. Solve the equation $y - 8 = 1$ for y.

4. Solve the equation $5b = 20$ for b.

5. Solve the equation $\frac{x}{4} = 10$ for x.

Worked Solutions

1. **No.** When you substitute 5 for a in the equation you have:

$$4 + (5) = 5$$
$$9 - 5$$

So $a = 5$ is not a solution to this equation.

2. **$x = 6$** Subtract 3 from both sides:

$$3 + x = 9$$
$$3 + x - 3 = 9 - 3$$
$$x = 6$$

Now insert 6 for x to check:

$$3 + (6) = 9$$
$$9 = 9 \checkmark$$

3. **$y = 9$**   Add 8 to both sides:

$$y - 8 = 1$$
$$y - 8 + 8 = 1 + 8$$
$$y = 9$$

Now substitute 9 for y to check:

$$(9) - 8 = 1$$
$$1 = 1 \checkmark$$

4. **$b = 4$**   Divide both sides by 5:

$$5b = 20$$
$$\frac{5b}{5} = \frac{20}{5}$$
$$\frac{\cancel{5}b}{\cancel{5}} = \frac{\overset{4}{\cancel{20}}}{\cancel{5}}$$
$$b = 4$$

Now substitute 4 for b to check:

$$5(4) = 20$$
$$20 = 20 \checkmark$$

5. **$x = 40$**   Multiply both sides by 4:

$$\frac{x}{4} = 10$$
$$4 \cdot \frac{x}{4} = 4 \cdot 10$$
$$\cancel{4} \cdot \frac{x}{\cancel{4}} = 4 \cdot 10$$
$$x = 40$$

Now substitute 40 for x to check:

$$\frac{40}{4} = 10$$
$$10 = 10 \checkmark$$

Combinations of Operations

Many times more than one step is required to solve an equation. The usual approach is to start working with the value that is farthest from the variable and work inward (some people like to think of it as peeling off layers of an onion). The goal is still to get the variable by itself on one side of the equation. In every case, you "undo" what's being done to the variable by doing the opposite—multiplication instead of division, addition instead of subtraction, and vice versa.

Examples and Problems

These problems show the answers and solutions.

1. Solve $3x + 6 = 18$ for x.

    answer: $x = 4$

    To solve, subtract 6 from both sides, getting $3x$ by itself:

    $$\begin{array}{rcl} 3x + 6 & = & 18 \\ -6 & & -6 \\ \hline 3x & = & 12 \end{array}$$

    Next, divide both sides by 3 to get x alone:

    $$\frac{3x}{3} = \frac{12}{3}$$

    $$\frac{\cancel{3}x}{\cancel{3}} = \frac{\cancel{12}^{4}}{\cancel{3}_{1}}$$

    $$x = 4$$

    To check, substitute 4 for x in the original equation:

    $$\begin{array}{rcl} 3(4) + 6 & = & 18 \\ 12 + 6 & = & 18 \\ 18 & = & 18 \quad \checkmark \end{array}$$

2. Solve $5y - 2 = 13$ for y.

    answer: $y = 3$

    First, add 2 to both sides:

    $$\begin{array}{rcl} 5y - 2 & = & 13 \\ +2 & & +2 \\ \hline 5y & = & 15 \end{array}$$

Next, divide both sides by 5:

$$\frac{5y}{5} = \frac{15}{5}$$

$$\frac{\cancel{5}y}{\cancel{5}} = \frac{\cancel{15}^{3}}{\cancel{5}_{1}}$$

$$y = 3$$

To check:

$$5(3) - 2 = 13$$
$$15 - 2 = 13$$
$$13 = 13 \quad \checkmark$$

3. Solve $\frac{x}{2} + 4 = 6$ for x.

   answer: $x = 4$

   Subtract 4 from both sides:

$$\frac{x}{2} + 4 = 6$$
$$\underline{\quad -4 \quad -4}$$
$$\frac{x}{2} \quad = 2$$

   Next, multiply each side by 2:

$$2 \cdot \frac{x}{2} = 2 \cdot 2$$
$$\cancel{2} \cdot \frac{x}{\cancel{2}} = 4$$
$$x = 4$$

   To check:

$$\frac{(4)}{2} + 4 = 6$$
$$2 + 4 = 6$$
$$6 = 6 \quad \checkmark$$

4. Solve $\frac{4b}{3} - 1 = 7$ for b.

   answer: $b = 6$

   Add 1 to each side:

$$\frac{4b}{3} - 1 = 7$$
$$\underline{\quad +1 \quad +1}$$
$$\frac{4b}{3} \quad = 8$$

Now multiply each side by $\frac{3}{4}$:

$$\frac{3}{4} \cdot \frac{4b}{3} = \frac{3}{4} \cdot 8$$

$$\frac{\cancel{3}}{\cancel{4}} \cdot \frac{\cancel{4}b}{\cancel{3}} = \frac{3}{\cancel{4}} \cdot \overset{2}{\cancel{8}}$$

$$b = 6$$

To check:

$$\frac{4(6)}{3} - 1 = 7$$

$$\frac{24}{3} - 1 = 7$$

$$8 - 1 = 7$$

$$7 = 7 \quad \checkmark$$

5. Solve $8(x - 3) = 16$ for x.

    answer: $x = 5$

    From the previous examples, you could get the idea that adding or subtracting is always the first step in these problems, but this time multiplication by 8 is the outermost thing being done to your variable, so undoing it is your first step:

$$\frac{1}{8} \cdot 8(x - 3) = \frac{1}{8} \cdot 16$$

$$\frac{1}{\cancel{8}} \cdot \cancel{8}(x - 3) = \frac{1}{\cancel{8}} \cdot \overset{2}{\cancel{16}}$$

$$x - 3 = 2$$

Now to finish up, add 3 to both sides:

$$\begin{array}{r} x - 3 = 2 \\ \underline{+3 \quad +3} \\ x \quad\;\; = 5 \end{array}$$

To check:

$$8[(5) - 3] = 16$$

$$8(2) = 16$$

$$16 = 16 \quad \checkmark$$

6. Solve $4x + 2 = x + 5$ for x.

    answer: $x = 1$

    Subtract 2 from both sides:

$$4x + 2 = x + 5$$

$$4x + 2 - 2 = x + 5 - 2$$

$$4x = x + 3$$

Subtract x from both sides (remember, you want all the x's on the same side):

$$4x = x + 3$$
$$4x - x = x + 3 - x$$
$$3x = 3$$

Divide both sides by 3:

$$3x = 3$$
$$\frac{3x}{3} = \frac{3}{3}$$
$$x = 1$$

To check:

$$4(1) + 2 = (1) + 5$$
$$4 + 2 = 6$$
$$6 = 6 \quad \checkmark$$

7. Solve $6y - 4 = 3y + 11$ for y.

answer: $y = 5$

Add 4 to each side:

$$6y - 4 = 3y + 11$$
$$\underline{+4 \qquad +4}$$
$$6y = 3y + 15$$

Subtract $3y$ from each side:

$$6y = 3y + 15$$
$$\underline{-3y \quad -3y}$$
$$3y = 15$$

Divide each side by 3:

$$3y = 15$$
$$\frac{3y}{3} = \frac{15}{3}$$
$$y = 5$$

To check:

$$6(5) - 4 = 3(5) + 11$$
$$30 - 4 = 15 + 11$$
$$26 = 26 \quad \checkmark$$

8. Solve $2a + 1 + 4 = 7 + 8$ for a.

   answer: $a = 5$

   Add the like terms on each side:

$$2a + 5 = 15$$

   Subtract 5 from both sides:

$$\begin{array}{r} 2a + 5 = 15 \\ \underline{-5 \quad -5} \\ 2a \quad = 10 \end{array}$$

   Next, divide both sides by 2:

$$\begin{array}{c} 2a = 10 \\ \dfrac{2a}{2} = \dfrac{10}{2} \\ a = 5 \end{array}$$

   To check:

$$\begin{array}{c} 2(5) + 1 + 4 = 7 + 8 \\ 10 + 1 + 4 = 15 \\ 15 = 15 \quad \checkmark \end{array}$$

Work Problems

Use these problems to give yourself additional practice.

1. Solve $4x - 6 = 8$ for x.

2. Solve $\dfrac{x}{6} + 1 = 4$ for x.

3. Solve $5(a - 4) = 3$ for a.

4. Solve $6n - 4 = 3 + 2n$ for n.

5. Solve $\dfrac{3y - 4}{2} = y + 1$ for y.

Worked Solutions

1. $x = \dfrac{7}{2}$ First add 6 to both sides:

$$\begin{array}{r} 4x - 6 = 8 \\ \underline{+6 \quad +6} \\ 4x \quad = 14 \end{array}$$

Then divide both sides by 4:

$$\frac{4x}{4} = \frac{14}{4}$$

$$\frac{\cancel{4}x}{\cancel{4}} = \frac{\overset{7}{\cancel{14}}}{\underset{2}{\cancel{4}}}$$

$$x = \frac{7}{2}$$

2. **$x = 18$**   First subtract 1 from both sides:

$$\frac{x}{6} + 1 = 4$$

$$\underline{\quad -1 \quad -1 \quad}$$

$$\frac{x}{6} \quad\;\; = 3$$

Then multiply both sides by 6:

$$6 \cdot \frac{x}{6} = 6 \cdot 3$$

$$\cancel{6} \cdot \frac{x}{\cancel{6}} = 6 \cdot 3$$

$$x = 18$$

3. **$a = 4\frac{3}{5}$**   First multiply both sides by $\frac{1}{5}$:

$$\frac{1}{5} \cdot 5(a - 4) = \frac{1}{5} \cdot 3$$

$$\frac{1}{\cancel{5}} \cdot \cancel{5}(a - 4) = \frac{3}{5}$$

$$a - 4 = \frac{3}{5}$$

Now add 4 to both sides:

$$a - 4 = \frac{3}{5}$$

$$\underline{+4 \quad +4 \qquad}$$

$$a \quad\;\; = 4\frac{3}{5}$$

4. **$n = \frac{7}{4}$**   First subtract $2n$ from both sides:

$$6n - 4 = 3 + 2n$$

$$\underline{-2n \qquad\quad -2n}$$

$$4n - 4 = 3$$

Now add 4 to both sides:

$$4n - 4 = 3$$
$$\underline{+4 \quad +4}$$
$$4n = 7$$

Finally divide both sides by 4:

$$\frac{4n}{4} = \frac{7}{4}$$

$$\frac{\cancel{4}n}{\cancel{4}} = \frac{7}{4}$$

$$n = \frac{7}{4}$$

5. **$y = 6$**   First multiply both sides by 2:

$$2 \cdot \frac{3y - 4}{2} = 2 \cdot (y + 1)$$

$$\cancel{2} \cdot \frac{3y - 4}{\cancel{2}} = 2y + 2$$

$$3y - 4 = 2y + 2$$

Then subtract $2y$ from both sides and add 4 to both sides:

$$3y - 4 = 2y + 2$$
$$\underline{-2y + 4 \quad -2y + 4}$$
$$y \quad = \quad 6$$

Appendix

Sequences

A **sequence** of numbers (also sometimes called a **progression**) is a list of numbers in a specific order, like the following:

1, 2, 3, 4, 5, 6, 7, ...

$8, 4, 2, 1, \frac{1}{2}, \frac{1}{4}, \frac{1}{8}, \dots$

1, 4, 1, 5, 9, 2, 6, ...

The numbers are called the terms of the sequence and are sometimes referred to as the first term, second term, and so forth. Sometimes, there is an obvious pattern to the terms, sometimes a less obvious pattern, or no pattern at all. Notice that, because sequences can go on and on (as indicated by the "..." at the end), often there's no way to write them out completely. Usually, we assume that any pattern that appears in the terms we see will be continued in the later terms, but whether that assumption is valid depends on the situation. For our purposes here, we'll count on patterns being continued.

Many different situations can be treated as sequences of numbers. Populations in successive years, heights of successive wave crests, and many other natural phenomena can all be seen this way. In this appendix, we briefly explore two simple patterns that sequences sometimes follow.

Arithmetic Sequences

A sequence in which the difference between any pair of successive terms is the same is called an **arithmetic sequence**.

Example Problems

These problems show the answers and solutions.

1. Is 2, 4, 6, 8, 10, ... an arithmetic sequence?

   answer: Yes.

   The difference between each pair of successive terms is always 2, so this is an arithmetic sequence.

2. Is 1, 2, 4, 7, 11, 16, 22, ... an arithmetic sequence?

   answer: No.

   The difference between the first and second terms is $2 - 1 = 1$, and the difference between the second and third terms is $4 - 2 = 2$, so the difference between successive terms is not always the same, and this is not an arithmetic sequence.

209

Geometric Sequences

Another common type of sequence is the **geometric sequence**, in which the quotient of each pair of successive terms is the same.

Example Problems

These problems show the answers and solutions.

1. Is 1, 3, 5, 7, 9, ... a geometric sequence?

    answer: No.

    The quotient of the first and second terms is $\frac{1}{3}$, and the quotient of the second and third terms is $\frac{3}{5}$, and these are not equivalent fractions.

2. Is 1, 2, 4, 8, 16, ... a geometric sequence?

    answer: Yes.

    The quotient of the first and second terms is $\frac{1}{2}$, and the quotient of the second and third terms is $\frac{2}{4}$, which reduces to $\frac{1}{2}$. Notice that even though the first two terms have the same ratio, that really isn't enough (think about the sequence in Example Problem 2, "Arithmetic Sequences"). In this case, however, every other successive pair has the same ratio, so this is a geometric sequence.

Predicting the Next Term of a Sequence

If you can determine that a sequence is a particular type, like arithmetic or geometric, then it is possible to predict what the next term will be (as long as the pattern continues).

Example Problems

These problems show the answers and solutions.

1. What is the next term in the arithmetic sequence 3, 6, 9, 12, ... ?

    answer: 15

    First, you figure out what the difference between successive terms is. In this case, $6 - 3 = 3$, so the difference between successive terms is 3. Then the term after 12 must be 3 more than 12, and $12 + 3 = 15$, so the next term will be 15.

2. What is the next term in the geometric sequence 1, 3, 9, 27, ... ?

    answer: 81

Because you know that this is a geometric sequence, this time look for the quotient of successive terms. The quotient of the first and second terms is $\frac{1}{3}$, so each term is $\frac{1}{3}$ of the following term. Your next term will be some number x, so that $27 = \frac{1}{3}x$.

You can solve this equation by multiplying both sides by 3:

$$3 \cdot 27 = 3 \cdot \frac{1}{3}x$$
$$81 = x$$

The next term in the sequence must be 81.

Work Problems

Use these problems to give yourself additional practice.

1. Is 1, 3, 5, 7, 9, ... an arithmetic sequence?

2. Is 1, 3, 6, 10, 15, ... an arithmetic sequence?

3. Is 5, 10, 20, 40, ... a geometric sequence?

4. Is 1, −1, 1, −1, 1, ... a geometric sequence?

5. Is 1, 4, 9, 16, 25, ... an arithmetic sequence, a geometric sequence, or neither?

6. Is 3, 3, 3, 3, 3, ... an arithmetic sequence, a geometric sequence, or neither?

7. What is the next term in the arithmetic sequence 5, 10, 15, 20, ... ?

8. What is the next term in the geometric sequence 3, $\frac{1}{3}$, $\frac{1}{9}$, $\frac{1}{27}$, ... ?

Worked Solutions

1. **Yes.** The difference between successive terms is always 2.

2. **No.** The difference between successive terms are not always the same. Actually, this sequence is called the triangular numbers because they give the total number of dots in a pattern of dots arranged in an equilateral triangle with a certain number of dots on each side. Think about the standard arrangement of bowling pins, for instance.

3. **Yes.** The quotient of successive terms is always $\frac{1}{2}$.

4. **Yes.** The quotient of successive terms is always −1.

5. **Neither.** Actually, this sequence is called the square numbers, in part because they are 1^2, 2^2, 3^2, 4^2, 5^2, and so forth, and in part for reasons like those suggested in the solution to Problem 2.

6. **Both.** The difference between successive terms is always zero, and the quotient of successive terms is always 1.

7. **25.** The difference between successive terms is always 5, so we have $20 + 5 = 25$.

8. $\frac{1}{81}$. The quotient between successive terms is always 3, so solve the equation $\frac{1}{27} = 3x$.

Glossary

acre   A unit of area in the English system equal to one 640th part of a square mile.

additive inverse   The opposite or negative of a number. The sum of any number and its additive inverse is 0.

algebraic expression   An expression in which numbers are represented by letters.

area   The measure of the region inside a figure in the plane; the number of unit squares it would take to cover a shape.

arithmetic sequence   A sequence in which the difference between consecutive terms is constant.

associative property   The way the numbers in an expression are grouped does not matter. Addition and multiplication have the associative property.

bar chart   A visual way of presenting several quantities with bars of different lengths.

base   A side of a polygon used for finding its area, or the length of that side.

centi-   A prefix used in the metric system to mean hundredths.

century   A unit of time equal to 100 years.

circle   A figure consisting of all points in the plane lying a fixed distance from a particular center point.

circle chart   A visual way of showing how a total amount, represented by a full circle, is broken down into parts, represented by wedges of the circle.

circumference   The distance around a circle.

closure property   A set has the closure property for a particular operation if all results of the operation fall into the original set.

common denominator   Two fractions must have the same denominator in order to be added or subtracted.

commutative property   The order of the numbers in an equation does not matter. Addition and multiplication have the commutative property.

complex fraction   Another name for a compound fraction.

composite number   Positive numbers that have more than two factors.

compound fraction   A fraction that contains another fraction in its numerator or denominator, or both.

counting numbers   Another name for the natural numbers 1, 2, 3, 4, ... l.

cube   A number that is equal to another number raised to the third power, or multiplied by itself and then by itself again.

cube root   A number that, when cubed, produces a given number.

cup   A unit of volume in the English system equal to eight ounces or one quarter of a quart.

day   A unit of time equal to 24 hours.

deca-   A prefix used in the metric system to mean 10.

decade   A unit of time equal to 10 years.

deci-   A prefix used in the metric system to mean tenths.

decimals   Numbers that extend the place value system by using digits to the right of a decimal point to represent tenths, hundredths, and so forth.

denominator   The number on the bottom in a fraction.

diameter   The length of the longest line that can be drawn from one point on a circle to another point on that same circle.

distributive property   The number on the outside of the parentheses is distributed to (multiplied by) each term on the inside of the parentheses. Multiplication distributes over addition or subtraction.

dividend   The number that is to be divided.

divisor   The number of parts into which something is to be divided.

English system   The system of measurement using units like feet, pounds, and gallons.

equilateral triangle   A triangle whose sides are all of equal length.

evaluate an expression   Find the numerical value of an algebraic expression by substituting numbers for the variables.

even numbers   Integers divisible by 2.

expanded notation   Writing out the place value of each digit.

exponent   A small raised number written to the right of another number, indicating the number of times that number should be multiplied by itself.

factors   Numbers that multiply together to get a product.

fluid ounce   A unit of volume in the English system equal to $\frac{1}{8}$ of a cup.

foot   A unit of length in the English system equal to 12 inches or $\frac{1}{3}$ of a yard.

fractions   Another name for rational numbers, numbers that can be written in the form $\frac{a}{b}$ for some integer a and natural number b.

gallon   A unit of volume in the English system equal to four quarts.

geometric sequence   A sequence in which the ratio of consecutive terms is constant.

gram   The basic unit of mass in the metric system.

hecto-   A prefix used in the metric system to mean hundreds.

height   The perpendicular distance from a line containing the base of a polygon to the opposite corner used in finding its area.

hour   A unit of time equal to 60 minutes or $\frac{1}{24}$ of a day.

identity element for addition   Zero. When zero is added to any number, it gives the original number.

identity element for multiplication   One. Any number multiplied by one gives the original number.

improper fraction   A fraction with a numerator larger than its denominator; any improper fraction can also be written as a mixed number.

inch   A unit of length in the English system equal to $\frac{1}{12}$ of a foot.

independent events   Two or more events that have no influence on the other's chances of occurring.

index   A small number written to the left of a radical to indicate what root should be taken.

integers   Counting numbers, their negatives, and zero.

irrational numbers   Numbers that cannot be written as fractions, such as $\sqrt{2}$ and π.

isosceles triangle   A triangle with two sides equal in length.

kilo-   A prefix used in the metric system to mean thousands.

LCD   An abbreviation for least common denominator.

least common denominator   The smallest common denominator that can be found for two or more fractions.

line graph   A visual way of presenting a series of quantities, with each piece of data plotted as a point and lines connecting successive points.

liter   The basic unit of volume in the metric system.

meter   The basic unit of length in the metric system.

metric system   A system of measurement using units like meters, grams, and liters.

mile   A unit of length in the English system equal to 5280 feet.

milli-   A prefix used in the metric system to mean thousandths.

minute   A unit of time equal to 60 seconds or $\frac{1}{60}$ of an hour.

mixed number   A whole number plus a proper fraction; any mixed number can also be written as an improper fraction.

multiplicative inverse   The reciprocal of a number. When a number is multiplied by its reciprocal, the answer is 1.

natural numbers   Counting numbers such as 1, 2, 3, 4, 5, ... l.

negative integers   Whole numbers less than zero, such as $-1, -2, -3, -4, ...$ l.

numerator   The number on top in a fraction.

odd numbers   Integers not divisible by 2, such as 1, 3, 5, 7, ... l.

origin   The position of zero on a number line.

ounce   A unit of weight in the English system equal to $\frac{1}{16}$ of a pound.

parallelogram   A quadrilateral with two pairs of opposite sides parallel to each other.

percent   A number of parts out of 100.

perfect cube   An integer that is the cube of another integer.

perfect square   A natural number that is the square of another natural number.

perimeter   The total distance around the outside of a polygon.

pie chart (also called a circle chart)   A circle divided into "wedges," which shows how a total amount is broken down into parts.

place value system   Each digit in a number receives a specific value based on its position.

polygon   A closed figure drawn in the plane with three or more sides.

positive numbers   The natural numbers.

pound   A unit of weight in the English system equal to 16 ounces.

power   An exponent, indicating how many times a number is to be multiplied by itself.

prime factorization   A factoring of a number into prime numbers.

prime number   A number that can only be divided by itself and one.

probability   The number of favorable events divided by the total number of possible events.

progression   A list of numbers in a particular order; another word for sequence.

proper fraction   A fraction whose numerator is less (in absolute value) than its denominator.

quadrilateral   A polygon with exactly four sides.

quart   A unit of volume in the English system equal to 32 fluid ounces or $\frac{1}{4}$ of a gallon.

quotient   The result of a division.

radical sign $\sqrt{}$ symbol, which when placed over a number, indicates that a root of that number is to be taken.

radius   The distance from the center to the edge of a circle.

rational numbers   Fractions. Numbers that can be written in the general form $\frac{a}{b}$, where a can be any integer and b can be any natural number.

reciprocal   The multiplicative inverse of a number, or one over that number.

rectangle   A parallelogram with all right angles.

reduced   A fraction is reduced if its numerator and denominator have no factors in common.

remainder   In a division problem, the amount left over.

right angle   A 90-degree angle, like one of the corners of a square.

right triangle   A triangle in which one of the angles is a right angle.

root   A number that, when raised to a certain power, produces a given number.

rounding off   A method of approximation achieved by adjusting a number to a particular place value.

scientific notation   A way of writing numbers as a number between 1 and 10 multiplied by 10 raised to some power, especially useful for very large or very small numbers.

second   A unit of time equal to $\frac{1}{60}$ of a minute.

sequence   A list of numbers in a particular order.

SI system   The metric system. SI is an abbreviation for the French term *Système International*.

solution   A value for the variable that makes an equation true.

solving a simple equation   The process of finding a solution to an equation.

square   A quadrilateral with four equal sides and four right angles.

square   To square a number means to multiply that number times itself.

square foot   A unit of area measure in the English system equal to that of a square that is one foot in length on each side.

square inch   A unit of area measure in the English system equal to that of a square that is one inch in length on each side.

square mile   A unit of area measure in the English system equal to that of a square that is one mile in length on each side.

square root   The number which, when squared, produces a given number.

ton   A unit of weight in the English system equal to 2000 pounds.

trapezoid   A quadrilateral with two sides parallel.

triangle   A polygon with exactly three sides.

variable   A letter that is used to represent a number, such as a, b, c, x, y, or z.

week   A unit of time equal to 7 days.

whole numbers   The set of numbers containing 0, 1, 2, 3, 4,...

year   A unit of time equal to approximately 365 days (only approximately, because of leap years).

Customized Full-Length Exam

Problems

1. Which of the following are natural numbers? $(-3, \frac{1}{2}, \sqrt{7}, 5)$

   Answer: 5.

   If you answered correctly, go to problem 3.
   If you answered incorrectly, go to problem 2.

2. Which of the following are natural numbers? $(-\frac{7}{3}, 2, \pi, 10)$

   Answer: 2 and 10.

   If you answered correctly, go to problem 3.
   If you answered incorrectly, go to Collections of Numbers on page 27.

3. Which of the following are rational numbers? $(-\frac{8}{5}, -1, 0, \frac{\pi}{3}, \sqrt{3}, 6.2)$

   Answer: All but $\frac{\pi}{3}$ and $\sqrt{3}$.

   If you answered correctly, go to problem 5.
   If you answered incorrectly, go to problem 4.

4. Which of the following are rational numbers? $(-7, -\frac{1}{2}, 2, \pi, \sqrt[3]{4})$

   Answer: All but π and $\sqrt[3]{4}$.

   If you answered correctly, go to problem 5.
   If you answered incorrectly, go to Collections of Numbers on page 27.

5. Simplify $3[20 - 5(1 + 2)]$.

   Answer: 15

   If you answered correctly, go to problem 7.
   If you answered incorrectly, go to problem 6.

6. Simplify $5[35 - 9(4 - 1)]$.

   Answer: 40

   If you answered correctly, go to problem 7.
   If you answered incorrectly, go to Grouping Symbols and Order of Operations on page 31.

7. Simplify $5^2 \times 2 + 10 \div 2 - (4 - 1)$.

   Answer: 52

If you answered correctly, go to problem 9.
If you answered incorrectly, go to problem 8.

8. Simplify $20 \div 2^2 + 2 \times 3 - (10 - 3)$.

   Answer: 4

If you answered correctly, go to problem 9.
If you answered incorrectly, go to Order of Operations on page 32.

9. Add $13 + 17 + 24$.

   Answer: 54

If you answered correctly, go to problem 11.
If you answered incorrectly, go to problem 10.

10. Add $31 + 12 + 29$.

    Answer: 72

If you answered correctly, go to problem 11.
If you answered incorrectly, go to Adding and Subtracting on page 37.

11. Subtract $134 - 85$.

    Answer: 49

If you answered correctly, go to problem 13.
If you answered incorrectly, go to problem 12.

12. Subtract $467 - 182$.

    Answer: 285

If you answered correctly, go to problem 13.
If you answered incorrectly, go to Adding and Subtracting on page 37.

13. Multiply 37×6.

    Answer: 222

If you answered correctly, go to problem 15.
If you answered incorrectly, go to problem 14.

14. Multiply 83×4.

    Answer: 332

If you answered correctly, go to problem 15.
If you answered incorrectly, go to Multiplying on page 39.

15. Multiply 341 × 39.

   Answer: 13,299

If you answered correctly, go to problem 17.
If you answered incorrectly, go to problem 16.

16. Multiply 57 × 219.

   Answer: 12,483

If you answered correctly, go to problem 17.
If you answered incorrectly, go to Multiplying on page 39.

17. Divide 105 ÷ 3.

   Answer: 35

If you answered correctly, go to problem 19.
If you answered incorrectly, go to problem 18.

18. Divide 335 ÷ 5.

   Answer: 67

If you answered correctly, go to problem 19.
If you answered incorrectly, go to Dividing on page 41.

19. Divide 1610 ÷ 35.

   Answer: 46

If you answered correctly, go to problem 21.
If you answered incorrectly, go to problem 20.

20. Divide 1961 ÷ 53.

   Answer: 37

If you answered correctly, go to problem 21.
If you answered incorrectly, go to Dividing on page 41.

21. Round 4515 to the nearest ten.

   Answer: 4520

If you answered correctly, go to problem 23.
If you answered incorrectly, go to problem 22.

22. Round 6781 to the nearest hundred.

   Answer: 6800

If you answered correctly, go to problem 23.
If you answered incorrectly, go to Rounding Off on page 45.

23.   What are the factors of 28?

   Answer: 1, 2, 4, 7, 14, and 28

If you answered correctly, go to problem 25.
If you answered incorrectly, go to problem 24.

24.   What are the factors of 45?

   Answer: 1, 3, 5, 9, 15, 45

If you answered correctly, go to problem 25.
If you answered incorrectly, go to Factoring on page 50.

25.   Write the fraction $\frac{10}{35}$ in reduced form.

   Answer: $\frac{2}{7}$

If you answered correctly, go to problem 27.
If you answered incorrectly, go to problem 26.

26.   Write the fraction $\frac{12}{18}$ in reduced form.

   Answer: $\frac{2}{3}$

If you answered correctly, go to problem 27.
If you answered incorrectly, go to Reducing Fractions on page 58.

27.   Change $\frac{19}{3}$ to a mixed number.

   Answer: $6\frac{1}{3}$

If you answered correctly, go to problem 29.
If you answered incorrectly, go to problem 28.

28.   Change $\frac{17}{5}$ to a mixed number.

   Answer: $3\frac{2}{5}$

If you answered correctly, go to problem 29.
If you answered incorrectly, go to Mixed Numbers on page 60.

29.   Change $5\frac{1}{2}$ to an improper fraction.

   Answer: $\frac{11}{2}$

If you answered correctly, go to problem 31.
If you answered incorrectly, go to problem 30.

30.   Change $2\frac{1}{6}$ to an improper fraction.

   Answer: $\frac{13}{6}$

If you answered correctly, go to problem 31.
If you answered incorrectly, go to Mixed Numbers on page 60.

31. Add $\frac{1}{2} + \frac{3}{5}$.

Answer: $\frac{11}{10}$

If you answered correctly, go to problem 33.
If you answered incorrectly, go to problem 32.

32. Add $\frac{5}{3} + \frac{3}{8}$.

Answer: $\frac{49}{24}$

If you answered correctly, go to problem 33.
If you answered incorrectly, go to Adding and Subtracting Fractions on page 62.

33. Subtract $\frac{5}{6} - \frac{1}{7}$.

Answer: $\frac{29}{42}$

If you answered correctly, go to problem 35.
If you answered incorrectly, go to problem 34.

34. Subtract $\frac{7}{8} - \frac{2}{3}$.

Answer: $\frac{5}{24}$

If you answered correctly, go to problem 35.
If you answered incorrectly, go to Adding and Subtracting Fractions on page 62.

35. Add $2\frac{1}{2} + 1\frac{1}{5}$.

Answer: $3\frac{7}{10}$

If you answered correctly, go to problem 37.
If you answered incorrectly, go to problem 36.

36. Add $3\frac{1}{3} + 5\frac{1}{6}$.

Answer: $8\frac{1}{2}$

If you answered correctly, go to problem 37.
If you answered incorrectly, go to Adding Mixed Numbers on page 65.

37. Add $5\frac{5}{6} + 2\frac{1}{2}$.

Answer: $8\frac{1}{3}$

If you answered correctly, go to problem 39.
If you answered incorrectly, go to problem 38.

38. Add $1\frac{3}{4} + 2\frac{1}{3}$.

Answer: $4\frac{1}{12}$

If you answered correctly, go to problem 39.
If you answered incorrectly, go to Adding Mixed Numbers on page 65.

39. Subtract $2\frac{5}{8} - 1\frac{1}{2}$.

> **Answer:** $1\frac{1}{8}$

If you answered correctly, go to problem 41.
If you answered incorrectly, go to problem 40.

40. Subtract $5\frac{1}{2} - 1\frac{1}{3}$.

> **Answer:** $4\frac{1}{6}$

If you answered correctly, go to problem 41.
If you answered incorrectly, go to Subtracting Mixed Numbers on page 66.

41. Subtract $4\frac{1}{3} - 1\frac{2}{5}$.

> **Answer:** $2\frac{14}{15}$

If you answered correctly, go to problem 43.
If you answered incorrectly, go to problem 42.

42. Subtract $6\frac{1}{6} - 2\frac{1}{3}$.

> **Answer:** $3\frac{5}{6}$

If you answered correctly, go to problem 43.
If you answered incorrectly, go to Subtracting Mixed Numbers on page 66.

43. Multiply $\frac{5}{3} \times \frac{2}{15}$.

> **Answer:** $\frac{2}{9}$

If you answered correctly, go to problem 45.
If you answered incorrectly, go to problem 44.

44. Multiply $\frac{3}{4} \times \frac{2}{5}$.

> **Answer:** $\frac{3}{10}$

If you answered correctly, go to problem 45.
If you answered incorrectly, go to Multiplying Fractions on page 68.

45. Multiply $3\frac{1}{4} \times 2\frac{2}{5}$.

> **Answer:** $7\frac{4}{5}$

If you answered correctly, go to problem 47.
If you answered incorrectly, go to problem 46.

46. Multiply $5\frac{1}{2} + 2\frac{3}{8}$.

> **Answer:** $13\frac{1}{16}$

If you answered correctly, go to problem 47.
If you answered incorrectly, go to Multiplying Mixed Numbers on page 69.

47. Divide $\frac{4}{5} \div \frac{1}{2}$.

Answer: $\frac{8}{5}$

If you answered correctly, go to problem 49.
If you answered incorrectly, go to problem 48.

48. Divide $\frac{6}{5} \div \frac{1}{3}$.

Answer: $\frac{18}{5}$

If you answered correctly, go to problem 49.
If you answered incorrectly, go to Dividing Fractions on page 71.

49. Divide $\frac{1}{6} \div \frac{2}{3}$.

Answer: $\frac{1}{4}$

If you answered correctly, go to problem 51.
If you answered incorrectly, go to problem 50.

50. Divide $\frac{3}{8} \div \frac{3}{4}$.

Answer: $\frac{1}{2}$

If you answered correctly, go to problem 51.
If you answered incorrectly, go to Dividing Fractions on page 71.

51. Divide $6\frac{1}{3} \div 1\frac{1}{4}$.

Answer: $\frac{76}{15}$

If you answered correctly, go to problem 53.
If you answered incorrectly, go to problem 52.

52. Divide $3\frac{1}{5} \div 2\frac{2}{3}$.

Answer: $\frac{6}{5}$

If you answered correctly, go to problem 53.
If you answered incorrectly, go to Dividing Mixed Numbers on page 71.

53. Simplify $\dfrac{4\frac{1}{3}}{2\frac{1}{2}}$.

Answer: $\frac{26}{15}$

If you answered correctly, go to problem 55.
If you answered incorrectly, go to problem 54.

54. Simplify $\dfrac{1\frac{1}{4}}{5\frac{1}{3}}$.

    Answer: $\dfrac{15}{64}$

    If you answered correctly, go to problem 55.
    If you answered incorrectly, go to Compound Fractions on page 73.

55. Simplify $\dfrac{\frac{1}{2}+\frac{1}{3}}{2-\frac{1}{4}}$.

    Answer: $\dfrac{10}{21}$

    If you answered correctly, go to problem 57.
    If you answered incorrectly, go to problem 56.

56. Simplify $\dfrac{\frac{1}{4}-\frac{1}{5}}{\frac{1}{6}-\frac{1}{7}}$.

    Answer: $\dfrac{21}{10}$

    If you answered correctly, go to problem 57.
    If you answered incorrectly, go to Compound Fractions on page 73.

57. Which is greater, 0.45 or 0.409?

    Answer: 0.45

    If you answered correctly, go to problem 59.
    If you answered incorrectly, go to problem 58.

58. Which is greater, 3.14159 or 3.1428?

    Answer: 3.1428

    If you answered correctly, go to problem 59.
    If you answered incorrectly, go to Comparing Decimals on page 76.

59. Add 5.2 + 3.6.

    Answer: 8.8

    If you answered correctly, go to problem 61.
    If you answered incorrectly, go to problem 60.

60. Add 4.1 + 9.3.

    Answer: 13.4

    If you answered correctly, go to problem 61.
    If you answered incorrectly, go to Adding Decimals on page 78.

61. Add 15.7 + 24.4.

Answer: 40.1

If you answered correctly, go to problem 63.
If you answered incorrectly, go to problem 62.

62. Add 48.9 + 32.4.

Answer: 81.3

If you answered correctly, go to problem 63.
If you answered incorrectly, go to Adding Decimals on page 78.

63. Subtract 9.6 − 3.2.

Answer: 6.4

If you answered correctly, go to problem 65.
If you answered incorrectly, go to problem 64.

64. Subtract 6.7 − 2.1.

Answer: 4.6

If you answered correctly, go to problem 65.
If you answered incorrectly, go to Subtracting Decimals on page 79.

65. Subtract 52.3 − 10.4.

Answer: 41.9

If you answered correctly, go to problem 67.
If you answered incorrectly, go to problem 66.

66. Subtract 75.5 − 8.9.

Answer: 66.6

If you answered correctly, go to problem 67.
If you answered incorrectly, go to Subtracting Decimals on page 79.

67. Multiply 3.2 × 7.5.

Answer: 24

If you answered correctly, go to problem 69.
If you answered incorrectly, go to problem 68.

68. Multiply 4.5 × 0.31.

Answer: 1.395

If you answered correctly, go to problem 69.
If you answered incorrectly, go to Multiplying Decimals on page 81.

69. Multiply 406.1×0.037.

Answer: 15.0257

If you answered correctly, go to problem 71.
If you answered incorrectly, go to problem 70.

70. Multiply 670×0.0025.

Answer: 1.675

If you answered correctly, go to problem 71.
If you answered incorrectly, go to Multiplying Decimals on page 81.

71. Divide $16.4 \div 4$.

Answer: 4.1

If you answered correctly, go to problem 73.
If you answered incorrectly, go to problem 72.

72. Divide $31.5 \div 5$.

Answer: 6.3

If you answered correctly, go to problem 73.
If you answered incorrectly, go to Dividing Decimals on page 84.

73. Divide $13.44 \div 2.4$.

Answer: 5.6

If you answered correctly, go to problem 75.
If you answered incorrectly, go to problem 74.

74. Divide $31.85 \div 9.1$.

Answer: 3.5

If you answered correctly, go to problem 75.
If you answered incorrectly, go to Dividing Decimals on page 84.

75. Divide $67.1 \div 0.4$.

Answer: 167.75

If you answered correctly, go to problem 77.
If you answered incorrectly, go to problem 76.

76. Divide $8.34 \div 0.5$.

Answer: 16.68

If you answered correctly, go to problem 77.
If you answered incorrectly, go to Dividing Decimals on page 84.

77. Divide $4.6 \div 0.6$.

Answer: $7.\overline{6}$

If you answered correctly, go to problem 79.
If you answered incorrectly, go to problem 78.

78. Divide $0.58 \div 0.9$.

Answer: $0.6\overline{4}$

If you answered correctly, go to problem 79.
If you answered incorrectly, go to Dividing Decimals on page 84.

79. Round 56.793 to the nearest whole number.

Answer: 57

If you answered correctly, go to problem 81.
If you answered incorrectly, go to problem 80.

80. Round 95.73 to the nearest whole number.

Answer: 96

If you answered correctly, go to problem 81.
If you answered incorrectly, go to Estimating on page 89.

81. Change $\frac{5}{8}$ to a decimal.

Answer: 0.625

If you answered correctly, go to problem 83.
If you answered incorrectly, go to problem 82.

82. Change $\frac{7}{12}$ to a decimal.

Answer: $0.58\overline{3}$

If you answered correctly, go to problem 83.
If you answered incorrectly, go to Changing Fractions to Decimals on page 90.

83. Change 0.45 to a fraction.

Answer: $\frac{9}{20}$

If you answered correctly, go to problem 85.
If you answered incorrectly, go to problem 84.

84. Change 0.62 to a fraction.

Answer: $\frac{31}{50}$

If you answered correctly, go to problem 85.
If you answered incorrectly, go to Changing Decimals to Fractions on page 93.

85. Change $0.\overline{16}$ to a fraction.

Answer: $\dfrac{16}{99}$

If you answered correctly, go to problem 87.
If you answered incorrectly, go to problem 86.

86. Change $0.91\overline{6}$ to a fraction.

Answer: $\dfrac{11}{12}$

If you answered correctly, go to problem 87.
If you answered incorrectly, go to Changing Decimals to Fractions on page 93.

87. Change 0.30 to a percent.

Answer: 30%

If you answered correctly, go to problem 89.
If you answered incorrectly, go to problem 88.

88. Change 0.75 to a percent.

Answer: 75%

If you answered correctly, go to problem 89.
If you answered incorrectly, go to Changing Decimals to Percents on page 97.

89. Change 8% to a decimal.

Answer: 0.08

If you answered correctly, go to problem 91.
If you answered incorrectly, go to problem 90.

90. Change 85% to a decimal.

Answer: 0.85

If you answered correctly, go to problem 91.
If you answered incorrectly, go to Changing Percents to Decimals on page 98.

91. Change $\dfrac{3}{4}$ to a percent.

Answer: 75%

If you answered correctly, go to problem 93.
If you answered incorrectly, go to problem 92.

92. Change $\dfrac{2}{5}$ to a percent.

Answer: 40%

If you answered correctly, go to problem 93.
If you answered incorrectly, go to Changing Fractions to Percents on page 99.

93.   Change 35% to a fraction.

   Answer: $\frac{7}{20}$

If you answered correctly, go to problem 95.
If you answered incorrectly, go to problem 94.

94.   Change 60% to a fraction.

   Answer: $\frac{3}{5}$

If you answered correctly, go to problem 95.
If you answered incorrectly, go to Changing Percents to Fractions on page 100.

95.   What is 20% of 50?

   Answer: 10

If you answered correctly, go to problem 97.
If you answered incorrectly, go to problem 96.

96.   What is 75% of 60?

   Answer: 45

If you answered correctly, go to problem 97.
If you answered incorrectly, go to Finding Percents of a Number on page 101.

97.   24 is what percent of 60?

   Answer: 40%

If you answered correctly, go to problem 99.
If you answered incorrectly, go to problem 98.

98.   45 is what percent of 80?

   Answer: 56.25%

If you answered correctly, go to problem 99.
If you answered incorrectly, go to Finding What Percent One Number Is of Another Number on page 103.

99.   20 is 40% of what number?

   Answer: 50

If you answered correctly, go to problem 101.
If you answered incorrectly, go to problem 100.

100.   45 is 15% of what number?

   Answer: 300

If you answered correctly, go to problem 101.
If you answered incorrectly, go to Finding a Number When a Percent of It Is Known on page 105.

101. What is the percent increase if a colony of 100 ants grows to 500 ants?

   Answer: 400%

If you answered correctly, go to problem 103.
If you answered incorrectly, go to problem 102.

102. What is the percent decrease if a shirt that cost $60 goes on sale for $45?

   Answer: 25%

If you answered correctly, go to problem 103.
If you answered incorrectly, go to Finding Percent Increase or Percent Decrease on page 110.

103. Add $-6 + -3$.

   Answer: -9

If you answered correctly, go to problem 105.
If you answered incorrectly, go to problem 104.

104. Add $-12 + -5$.

   Answer: -17

If you answered correctly, go to problem 105.
If you answered incorrectly, go to Adding Integers on page 113.

105. Add $-4 + 7$.

   Answer: 3

If you answered correctly, go to problem 107.
If you answered incorrectly, go to problem 106.

106. Add $6 + -3$.

   Answer: 3

If you answered correctly, go to problem 107.
If you answered incorrectly, go to Adding Integers on page 113.

107. Add $-7 + 4$.

   Answer: -3

If you answered correctly, go to problem 109.
If you answered incorrectly, go to problem 108.

108. Add $15 + - 17$.

   Answer: -2

If you answered correctly, go to problem 109.
If you answered incorrectly, go to Adding Integers on page 113.

109. Subtract $(-12) - (-4)$.

Answer: -8

If you answered correctly, go to problem 111.
If you answered incorrectly, go to problem 110.

110. Subtract $(+10) - (-5)$.

Answer: 15

If you answered correctly, go to problem 111.
If you answered incorrectly, go to Subtracting Integers on page 115.

111. Subtract $12 - (14 - 3 + 2)$.

Answer: -1

If you answered correctly, go to problem 113.
If you answered incorrectly, go to problem 112.

112. Subtract $-8 - (5 + 3 - 1)$.

Answer: -15

If you answered correctly, go to problem 113.
If you answered incorrectly, go to Minus Preceding Grouping Symbols on page 117.

113. Multiply $(-2) \times (5)$.

Answer: -10

If you answered correctly, go to problem 115.
If you answered incorrectly, go to problem 114.

114. Multiply $(3) \times (-6)$.

Answer: -18

If you answered correctly, go to problem 115.
If you answered incorrectly, go to Multiplying and Dividing Integers on page 118.

115. Multiply $(-7) \times (-3)$.

Answer: 21

If you answered correctly, go to problem 117.
If you answered incorrectly, go to problem 116.

116. Multiply $(-2) \times (-5)$.

Answer: 10

If you answered correctly, go to problem 117.
If you answered incorrectly, go to Multiplying and Dividing Integers on page 118.

117. Divide $(-10) \div (2)$.

Answer: -5

If you answered correctly, go to problem 119.
If you answered incorrectly, go to problem 118.

118. Divide $(-12) \div (6)$.

Answer: -2

If you answered correctly, go to problem 119.
If you answered incorrectly, go to Multiplying and Dividing Integers on page 118.

119. Divide $(-18) \div (-3)$.

Answer: 6

If you answered correctly, go to problem 121.
If you answered incorrectly, go to problem 120.

120. Divide $(-15) \div (-5)$.

Answer: 3

If you answered correctly, go to problem 121.
If you answered incorrectly, go to Multiplying and Dividing Integers on page 118.

121. Multiply $(-3) \times (-5) \times (2) \times (4) \times (-2)$.

Answer: -240

If you answered correctly, go to problem 123.
If you answered incorrectly, go to problem 122.

122. Multiply $(-5) \times (2) \times (-3) \times (3)$.

Answer: 90

If you answered correctly, go to problem 123.
If you answered incorrectly, go to Multiplying and Dividing Integers on page 118.

123. Find the absolute value $|-3|$.

Answer: 3

If you answered correctly, go to problem 125.
If you answered incorrectly, go to problem 124.

124. Find the absolute value $|-8|$.

Answer: 8

If you answered correctly, go to problem 125.
If you answered incorrectly, go to Absolute Value on page 120.

125. Find the value of $4 - |-2|$.

      Answer: 2

If you answered correctly, go to problem 127.
If you answered incorrectly, go to problem 126.

126. Find the value of $7 - |-5|$.

      Answer: 2

If you answered correctly, go to problem 127.
If you answered incorrectly, go to Absolute Value on page 120.

127. Add $\dfrac{-5}{6} + \dfrac{1}{2}$.

      Answer: $\dfrac{-1}{3}$

If you answered correctly, go to problem 129.
If you answered incorrectly, go to problem 128.

128. Add $\dfrac{5}{8} + \dfrac{-2}{3}$.

      Answer: $\dfrac{-1}{24}$

If you answered correctly, go to problem 129.
If you answered incorrectly, go to Adding Positive and Negative Fractions on page 122.

129. Subtract $\dfrac{-1}{3} - \dfrac{1}{4}$.

      Answer: $\dfrac{-7}{12}$

If you answered correctly, go to problem 131.
If you answered incorrectly, go to problem 130.

130. Subtract $\dfrac{3}{4} - \dfrac{-1}{3}$.

      Answer: $\dfrac{13}{12}$

If you answered correctly, go to problem 131.
If you answered incorrectly, go to Subtracting Positive and Negative Fractions on page 126.

131. Multiply $\dfrac{2}{7} \times \dfrac{-3}{4}$.

      Answer: $\dfrac{-3}{14}$

If you answered correctly, go to problem 133.
If you answered incorrectly, go to problem 132.

132. Multiply $\dfrac{-4}{5} \times \dfrac{-3}{8}$.

      Answer: $\dfrac{3}{10}$

If you answered correctly, go to problem 133.
If you answered incorrectly, go to Multiplying Positive and Negative Fractions on page 128.

133. Divide $\frac{-2}{5} \div \frac{-6}{11}$.

Answer: $\frac{11}{15}$

If you answered correctly, go to problem 135.
If you answered incorrectly, go to problem 134.

134. Divide $\frac{-5}{7} \div \frac{2}{3}$.

Answer: $\frac{-15}{14}$

If you answered correctly, go to problem 135.
If you answered incorrectly, go to Dividing Positive and Negative Fractions on page 131.

135. Expand 2^3.

Answer: 8

If you answered correctly, go to problem 137.
If you answered incorrectly, go to problem 136.

136. Expand 3^4.

Answer: 81

If you answered correctly, go to problem 137.
If you answered incorrectly, go to Exponents on page 135.

137. Expand 5^{-2}.

Answer: $\frac{1}{25}$

If you answered correctly, go to problem 139.
If you answered incorrectly, go to problem 138.

138. Expand 6^{-2}.

Answer: $\frac{1}{36}$

If you answered correctly, go to problem 139.
If you answered incorrectly, go to Negative Exponents on page 136.

139. Multiply $5^{-2} \times 5^3$.

Answer: 5

If you answered correctly, go to problem 141.
If you answered incorrectly, go to problem 140.

140. Multiply $7^6 \times 7^{-4}$.

Answer: 49

If you answered correctly, go to problem 141.
If you answered incorrectly, go to Operations with Powers and Exponents on page 138.

141. Divide $9^3 \div 9^2$.

Answer: 9

If you answered correctly, go to problem 143.
If you answered incorrectly, go to problem 142.

142. Divide $3^{-3} \div 3^{-2}$.

Answer: $\frac{1}{3}$

If you answered correctly, go to problem 143.
If you answered incorrectly, go to Operations with Powers and Exponents on page 138.

143. Write $(2^5)^4$ with a single exponent.

Answer: 2^{20}

If you answered correctly, go to problem 145.
If you answered incorrectly, go to problem 144.

144. Write $(3^2)^4$ with a single exponent.

Answer: 3^8

If you answered correctly, go to problem 145.
If you answered incorrectly, go to More Operations with Powers and Exponents on page 140.

145. What is $\sqrt{25}$?

Answer: 5

If you answered correctly, go to problem 147.
If you answered incorrectly, go to problem 146.

146. What is $\sqrt{81}$?

Answer: 9

If you answered correctly, go to problem 147.
If you answered incorrectly, go to Square Roots on page 142.

147. What is $\sqrt[3]{8}$?

Answer: 2

If you answered correctly, go to problem 149.
If you answered incorrectly, go to problem 148.

148. What is $\sqrt[3]{27}$?

Answer: 3

If you answered correctly, go to problem 149.
If you answered incorrectly, go to Cube Roots on page 143.

149. Simplify $\sqrt{50}$.

Answer: $5\sqrt{2}$

If you answered correctly, go to problem 151.
If you answered incorrectly, go to problem 150.

150. Simplify $\sqrt{72}$.

Answer: $6\sqrt{2}$

If you answered correctly, go to problem 151.
If you answered incorrectly, go to Simplifying Square Roots on page 144.

151. Write 0.0045 in scientific notation.

Answer: 4.5×10^{-3}

If you answered correctly, go to problem 153.
If you answered incorrectly, go to problem 152.

152. Write 72,000,000 in scientific notation.

Answer: 7.2×10^{7}

If you answered correctly, go to problem 153.
If you answered incorrectly, go to Scientific Notation on page 149.

153. Multiply $(3.4 \times 10^{6}) \times (2.0 \times 10^{-4})$.

Answer: 6.8×10^{2}

If you answered correctly, go to problem 155.
If you answered incorrectly, go to problem 154.

154. Multiply $(1.8 \times 10^{8}) \times (3.0 \times 10^{3})$.

Answer: 5.4×10^{11}

If you answered correctly, go to problem 155.
If you answered incorrectly, go to Multiplying in Scientific Notation on page 150.

155. Divide $(1.8 \times 10^{8}) \div (3.0 \times 10^{3})$.

Answer: 6.0×10^{4}

If you answered correctly, go to problem 157.
If you answered incorrectly, go to problem 156.

156. Divide $(3.4 \times 10^{6}) \div (2.0 \times 10^{-4})$.

Answer: 1.7×10^{10}

If you answered correctly, go to problem 157.
If you answered incorrectly, go to Dividing in Scientific Notation on page 152.

157. What is the area of a triangle with a base of 6 inches and a height of 10 inches?

 Answer: 30 square inches

If you answered correctly, go to problem 159.
If you answered incorrectly, go to problem 158.

158. What is the area of a parallelogram with a base of 3 centimeters and height of 8 centimeters?

 Answer: 24 square centimeters.

If you answered correctly, go to problem 159.
If you answered incorrectly, go to Measurement of Basic Figures on page 163.

159. What is the perimeter of a circle with a radius of 3 feet?

 Answer: 6π

If you answered correctly, go to problem 161.
If you answered incorrectly, go to problem 160.

160. What is the area of a circle with a diameter of 10 centimeters?

 Answer: 25π

If you answered correctly, go to problem 161.
If you answered incorrectly, go to Circles on page 168.

161. If a standard die is tossed, what is the probability of rolling a 4 or less?

 Answer: $\frac{2}{3}$

If you answered correctly, go to problem 163.
If you answered incorrectly, go to problem 162.

162. If a marble is drawn at random from a jar containing one red marble, one green marble, and five black marbles, what is the probability of drawing a marble that is not black?

 Answer: $\frac{2}{7}$

If you answered correctly, go to problem 163.
If you answered incorrectly, go to Probability on page 179.

163. What are the mean, median, and mode of 5, 7, 2, 5, and 3?

 Answer: mean 4.4, median 5, mode 5.

If you answered correctly, go to problem 165.
If you answered incorrectly, go to problem 164.

164. What are the mean, median, and mode of 13, 12, 13, and 10?

    Answer: mean 12, median 12.5, mode 13.

    If you answered correctly, go to problem 165.
    If you answered incorrectly, go to Statistics on page 186.

165. Give an algebraic expression for 3 less than a number (use x to represent the number).

    Answer: $x - 3$

    If you answered correctly, go to problem 167.
    If you answered incorrectly, go to problem 166.

166. Give an algebraic expression for twice a number.

    Answer: $2x$

    If you answered correctly, go to problem 167.
    If you answered incorrectly, go to Algebraic Expressions on page 189.

167. Evaluate $5x - 2$ for $x = 3$.

    Answer: 13

    If you answered correctly, go to problem 169.
    If you answered incorrectly, go to problem 168.

168. Evaluate $3x + 1$ for $x = 7$.

    Answer: 22

    If you answered correctly, go to problem 169.
    If you answered incorrectly, go to Evaluating Expressions on page 191.

169. Solve $4x - 2 = 18$ for x.

    Answer: $x = 5$

    If you answered correctly, you're done!
    If you answered incorrectly, go to problem 170.

170. Solve $\frac{y}{3} + 4 = 7$ for y.

    Answer: $y = 9$

    If you answered correctly, go to bed now, you've done enough!
    If you answered incorrectly, go to Solving Simple Equations on page 194.

Index

Wiley Publishing, Inc.
End-User License Agreement

READ THIS. You should carefully read these terms and conditions before opening the software packet(s) included with this book "Book". This is a license agreement "Agreement" between you and Wiley Publishing, Inc. "WPI". By opening the accompanying software packet(s), you acknowledge that you have read and accept the following terms and conditions. If you do not agree and do not want to be bound by such terms and conditions, promptly return the Book and the unopened software packet(s) to the place you obtained them for a full refund.

1. **License Grant.** WPI grants to you (either an individual or entity) a nonexclusive license to use one copy of the enclosed software program(s) (collectively, the "Software") solely for your own personal or business purposes on a single computer (whether a standard computer or a workstation component of a multi-user network). The Software is in use on a computer when it is loaded into temporary memory (RAM) or installed into permanent memory (hard disk, CD-ROM, or other storage device). WPI reserves all rights not expressly granted herein.

2. **Ownership.** WPI is the owner of all right, title, and interest, including copyright, in and to the compilation of the Software recorded on the physical packet included with this Book "Software Media". Copyright to the individual programs recorded on the Software Media is owned by the author or other authorized copyright owner of each program. Ownership of the Software and all proprietary rights relating thereto remain with WPI and its licensers.

3. **Restrictions on Use and Transfer.**

   (a) You may only (i) make one copy of the Software for backup or archival purposes, or (ii) transfer the Software to a single hard disk, provided that you keep the original for backup or archival purposes. You may not (i) rent or lease the Software, (ii) copy or reproduce the Software through a LAN or other network system or through any computer subscriber system or bulletin-board system, or (iii) modify, adapt, or create derivative works based on the Software.

   (b) You may not reverse engineer, decompile, or disassemble the Software. You may transfer the Software and user documentation on a permanent basis, provided that the transferee agrees to accept the terms and conditions of this Agreement and you retain no copies. If the Software is an update or has been updated, any transfer must include the most recent update and all prior versions.

4. **Restrictions on Use of Individual Programs.** You must follow the individual requirements and restrictions detailed for each individual program on the Software Media. These limitations are also contained in the individual license agreements recorded on the Software Media. These limitations may include a requirement that after using the program for a specified period of time, the user must pay a registration fee or discontinue use. By opening the Software packet(s), you agree to abide by the licenses and restrictions for these individual programs that are detailed on the Software Media. None of the material on this Software Media or listed in this Book may ever be redistributed, in original or modified form, for commercial purposes.

5. **Limited Warranty.**

   (a) WPI warrants that the Software and Software Media are free from defects in materials and workmanship under normal use for a period of sixty (60) days from the date of purchase of this Book. If WPI receives notification within the warranty period of defects in materials or workmanship, WPI will replace the defective Software Media.

(b) WPI AND THE AUTHOR(S) OF THE BOOK DISCLAIM ALL OTHER WARRANTIES, EXPRESS OR IMPLIED, INCLUDING WITHOUT LIMITATION IMPLIED WARRANTIES OF MER-CHANTABILITY AND FITNESS FOR A PARTICULAR PURPOSE, WITH RESPECT TO THE SOFTWARE, THE PROGRAMS, THE SOURCE CODE CONTAINED THEREIN, AND/OR THE TECHNIQUES DESCRIBED IN THIS BOOK. WPI DOES NOT WARRANT THAT THE FUNC-TIONS CONTAINED IN THE SOFTWARE WILL MEET YOUR REQUIREMENTS OR THAT THE OPERATION OF THE SOFTWARE WILL BE ERROR FREE.

(c) This limited warranty gives you specific legal rights, and you may have other rights that vary from jurisdiction to jurisdiction.

6. Remedies.

(a) WPI's entire liability and your exclusive remedy for defects in materials and workmanship shall be limited to replacement of the Software Media, which may be returned to WPI with a copy of your receipt at the following address: Software Media Fulfillment Department, Attn.: *CliffsNotes® Basic Math and Pre-Algebra Practice Pack,* Wiley Publishing, Inc., 10475 Crosspoint Blvd., Indianapolis, IN 46256, or call 1-877-762-2974. Please allow four to six weeks for delivery. This Limited Warranty is void if failure of the Software Media has re-sulted from accident, abuse, or misapplication. Any replacement Software Media will be warranted for the remainder of the original warranty period or thirty (30) days, whichever is longer.

(b) In no event shall WPI or the author be liable for any damages whatsoever (including with-out limitation damages for loss of business profits, business interruption, loss of business information, or any other pecuniary loss) arising from the use of or inability to use the Book or the Software, even if WPI has been advised of the possibility of such damages.

(c) Because some jurisdictions do not allow the exclusion or limitation of liability for conse-quential or incidental damages, the above limitation or exclusion may not apply to you.

7. U.S. Government Restricted Rights. Use, duplication, or disclosure of the Software for or on behalf of the United States of America, its agencies and/or instrumentalities "U.S. Government" is subject to restrictions as stated in paragraph (c)(1)(ii) of the Rights in Technical Data and Computer Software clause of DFARS 252.227-7013, or subparagraphs (c) (1) and (2) of the Commercial Computer Software - Restricted Rights clause at FAR 52.227-19, and in similar clauses in the NASA FAR supplement, as applicable.

8. General. This Agreement constitutes the entire understanding of the parties and revokes and supersedes all prior agreements, oral or written, between them and may not be modi-fied or amended except in a writing signed by both parties hereto that specifically refers to this Agreement. This Agreement shall take precedence over any other documents that may be in conflict herewith. If any one or more provisions contained in this Agreement are held by any court or tribunal to be invalid, illegal, or otherwise unenforceable, each and every other provision shall remain in full force and effect.

Practice *does* make perfect with CliffsNotes®!

CliffsNotes Practice Packs help you master all of your key subjects with hundreds of practice problems and their solutions.

Bonus CD includes hundreds of additional practice problems!

CliffsNotes
Basic Math and Pre-Algebra
CD includes 600 practice problems
Practice Pack
The learn-by-doing way to master Basic Math and Pre-Algebra

▶ Pretest that pinpoints what you need to study most
▶ Clear, concise reviews of every topic
▶ Targeted practice problems in every chapter with solutions and explanations
▶ Customized full-length practice test that adapts to your skill level

Jonathan J. White, Scott Searcy, Teri Stimmel, and Danielle Lutz

978-0-470-53349-9

CliffsNotes
Algebra I
CD includes 500 practice problems
Practice Pack
The learn-by-doing way to master Algebra I

▶ Pretest that pinpoints what you need to study most
▶ Clear, concise reviews of every topic
▶ Targeted practice problems in every chapter with solutions and explanations
▶ Customized full-length practice test that adapts to your skill level

Mary Jane Sterling

978-0-470-49596-4

CliffsNotes
Algebra II
CD includes 500 practice problems
Practice Pack
The learn-by-doing way to master Algebra II

▶ Pretest that pinpoints what you need to study most
▶ Clear, concise reviews of every topic
▶ Targeted practice problems in every chapter with solutions and explanations
▶ Customized full-length practice test that adapts to your skill level

Mary Jane Sterling

978-0-470-49597-1

CliffsNotes
Geometry
CD includes 400 practice problems
Practice Pack
The learn-by-doing way to master Geometry

▶ Pretest that pinpoints what you need to study most
▶ Clear, concise reviews of every topic
▶ Targeted practice problems in every chapter with solutions and explanations
▶ Customized full-length practice test that adapts to your skill level

David A. Herzog

978-0-470-48869-0

CliffsNotes
Chemistry
CD includes 500 practice problems
Practice Pack
The learn-by-doing way to master Chemistry

▶ Pretest that pinpoints what you need to study most
▶ Clear, concise reviews of every topic
▶ Targeted practice problems in every chapter with solutions and explanations
▶ Customized full-length practice test that adapts to your skill level

Charles Henrickson

978-0-470-49595-7

CliffsNotes
English Grammar
CD includes 400 practice problems
Practice Pack
The learn-by-doing way to master English Grammar

▶ Pretest that pinpoints what you need to study most
▶ Clear, concise reviews of every topic
▶ Targeted practice problems in every chapter with solutions and explanations
▶ Customized full-length practice test that adapts to your skill level

Jeffrey G. Coghill and Stacy Magedanz

978-0-470-49639-8

Available wherever books are sold or visit us at CliffsNotes.com®

CliffsNotes and CliffsNotes.com are registered trademarks of Wiley Publishing, Inc.

CliffsNotes
A Branded Imprint of ⊕WIL
Now you know.